BIM Development and Trends in Developing Countries: Case Studies

Authored By

John Rogers

Roamef
UK

Heap-Yih Chong

School of Built Environment
Curtin University
Australia

Christopher Preece

Razak School of Engineering & Advanced Technology
Universiti Teknologi Malaysia
Malaysia

Chai Chai Lim

Faculty of Science and Engineering
Universiti Tunku Abdul Rahman
Malaysia

&

Himal Suranga Jayasena

Faculty of Architecture
University of Moratuwa
Sri Lanka

CONTENTS

CHAPTERS

Bio- Sketch of Authors

Dr. Rogers obtained his DBA in 2013 from Southern Cross University, Australia. Currently, he serves as MD for Roamef, which is a leading BIM-solutions provider in the UK and ASEAN. The company is working as a partner with the British and Malaysian governments.

Dr. Chong is a Senior Lecturer in the Department of Construction Management, School of Built Environment, Curtin University, Australia. His current research areas are BIM, ICT, Construction Project Management, Sustainability and Contract Administration.

Professor Dr. Preece is a Director of UTM International, Kuala Lumpur Campus and also Professor of International Construction Business, Universiti Teknologi Malaysia. He also serves as CIOB Construction Ambassador.

Assistant Professor Lim is a construction expert. He had been working in construction industry for more than 25 years in the construction industry and 10 years in lecturing, and served numerous roles, like general manager, contract manager, quantity surveyor in various companies including a listed contractor firm in Malaysia.

Suranga is a Senior Lecturer in the Department of Building Economics, Faculty of Architecture, University of Moratuwa, Sri Lanka and is the founder of BIMLab Network research group. His current research interests are in BIM, IPD, Building Economics and Contract Management.

Salient Features

Building Information Modelling (BIM) has been effectively adopted and benefited in numerous projects across the globe, particularly in developed countries. This book aims to address the philosophies, applications and trends of BIM for its better adoption in developing countries. Two case studies were selected in Malaysia and Sri Lanka. This book provides useful insights on BIM evolution and usage. It is the first in the market that is reporting the developing countries' scenarios towards an emerging BIM technology. The book would create a great impact and new insights into future uptake and development of BIM worldwide.

BIOGRAPHY

John Rogers had extensive experience in construction before migrating to the IT sector to help digitalise the telecomms, aerospace and banking industries. This digitalisation is now happening to the construction industry as BIM. To support this process in Malaysia, John co-founded the Malaysian chapter of building SMART, the independent, global authority for BIM. His Doctorate specifically addressed the opportunities from and barriers to the adoption of BIM in Malaysia. John works in the commercial sector helping large organisations shape corporate strategy to maximise competitive advantage potential and adopt BIM into their operational environment. He is a Research Fellow at City University of Science and Technology, with specific focus on high-rise, volumetric and modular construction utilising BIM and advanced modelling techniques.

Dr. Chong is a Senior Lecturer in Department of Construction Management, School of Built Environment, Curtin University, Australia. He brings with him a sound understanding of the construction sector and has a strong track record of coherent research outputs in the top ranked refereed journal articles, and successfully secured a number of competitive research grants from funding bodies in Malaysia, Japan and Australia. His current research areas are BIM, ICT, Construction Project Management, Sustainability and Contract Administration.

Professor Dr. Preece is a Professor of International Construction Business at Universiti Teknologi Malaysia (UTM), at the International Campus in Kuala Lumpur. His specialist research field is in the area of business management in the engineering sectors and is currently directing research in the areas of sustainability, value management in construction, marketing green services, asset management and BIM.

C C Lim is an Assistant Professor at the Universiti Tunku Abdul Rahman. He has twenty five years of experience in the Civil Engineering and Building Industry and specialising in construction administration and project management. Later on he entered the academic environment and lecturing on construction related subjects for the past ten years and taking a special interest in BIM knowledge development and its adoption by the construction industry players.

Suranga is a Senior Lecturer in the Department of Building Economics, Faculty of Architecture, University of Moratuwa, Sri Lanka and is the founder of BIMLab Network research group. His current research interests are in BIM, IPD, Building Economics and Contract Management. He is also the chair of CIOB Sri Lankan Centre.

PREFACE

Building Information Modelling (BIM) has been effectively adopted and benefited in numerous projects across the globe, particularly in developed countries. This book aims to address the philosophies, applications and trends of BIM for its better adoption in developing countries. Two case studies were selected in Malaysia and Sri Lanka. The flow of the book begins with the introduction and background of the Malaysian and Sri Lankan construction industry and follows by the critical review of BIM's philosophies, development and applications in different stages of the project as well as some discussions on certain BIM software. Subsequently, the way forward of BIM was articulated from the two perspectives, namely, academia and industry/BIM practitioner. The perspectives reveal that a different management and availability of library data are imperative for effective adoption of BIM, and integrations with other technologies are required for BIM in the near future, such as cloud computing, mobile devices, *etc*. The case studies highlight that the proper guidance of BIM software/technicality and adequacy of governmental supports were the main challenges at the moment. The book provides useful insight on BIM evolution and usage in Malaysia and Sri Lanka. It can also serves as a practical reference for related studies and practice in developing countries. In conclusion, readers will appreciate the fundamental approach of the discussion on the growing BIM body of knowledge, philosophies and trends.

ACKNOWLEDGEMENTS

Declared None.

CONFLICT OF INTEREST

The authors confirm that this book contents have no conflict of interest.

John Rogers
Roamef
UK

Heap-Yih Chong
School of Built Environment
Curtin University
Australia
E-mail: heap-yih.chong@curtin.edu.au

Christopher Preece
Razak School of Engineering & Advanced Technology
Universiti Teknologi Malaysia
Malaysia

Chai Chai Lim
Faculty of Science and Engineering
Universiti Tunku Abdul Rahman
Malaysia

Himal Suranga Jayasena
Faculty of Architecture
University of Moratuwa
Sri Lanka

CHAPTER 1

Introduction

Abstract: Building Information Technology (BIM) has proven effective and useful across the sectors in the built environment. This is critical to promote this emerging technology, particularly to developing countries. Malaysia and Sri Lanka have been selected for this book to highlight different levels of BIM position and to explain their different backgrounds and the need for BIM development. It provides a clear flow of message and understanding about relevant developments and activities in the construction sector. The discussions made towards the countries can be referred by other developing countries. The findings are able to provide new insights into future uptake and development of BIM worldwide.

Building Information Technology (BIM) can be described as an emerging technology, an advanced information technology tool or even a collaborative digital working platform. BIM is now and will be the future all sectors in the built environment as to embrace the impact of globalization and advantages of competitiveness.

The adoption and use of BIM have been actively promoted and implemented in developed countries. For instance, the recent initiative by United Kingdom has extended and mandated the BIM maturity to Level 2 in 2016 for all central government projects that exceed certain amount of contract sum. Meanwhile, other countries have also started certain mandatories for BIM digital files submission in order for local authorities' approval, such as Singapore and Taiwan.

On the other hand, developing countries are aware of BIM wave of development. Numerous conferences and promotions have been conducted for BIM in the respective countries, namely Malaysia and Sri Lanka. The countries have been selected for this book to highlight different levels of BIM position as per their Human Development Index in the developing countries' development status. It would explain their different backgrounds and the need for BIM development. The discussions made towards the countries might not represent the whole developing countries, but are able to provide insightful messages for the rest of developing countries.

Therefore, we organize the flow of the book by articulating the background of construction industries based on the available sources from the respective countries. It provides certain understandings about the history of countries and relevant developments and activities in the construction sector. Subsequently, we

John Rogers, Heap-Yih Chong, Christopher Preece, Chai Chai Lim and Himal Suranga Jayasena

discuss the current development and evolution of BIM at length. The mixed perspectives from industry and academia are presented for the future outlook and the way forward for BIM research and implementation. Last but not least, we adopt case study approach and discuss the contemporary BIM development in Malaysia and Sri Lanka.

BIM has been proven and benefited in numerous projects according to literatures. This book has addressed the philosophies, applications and trends of BIM for its better adoption in developing countries, based on the case studies in Malaysia and Sri Lanka. This is the first or a very rare in the market that is reporting the developing countries' scenarios towards an emerging BIM technology. It would provide new insights into future uptake and development of BIM worldwide.

CHAPTER 2

Malaysian Construction Industry

Abstract: Malaysian construction industry contributes significantly to the economy. Several professional bodies and institutions are governing the practice in the industry. Numerous policies have been prescribed that are intended to address chronic, systemic weaknesses within the construction sector. Generally, these follow global trends and lead the world in application. Generally too, their application is on public sector projects where the sophisticated client can leverage demand side insistence that new techniques and systems are employed. On the mid-range, conventionally procured, private sector projects that are the single largest market sector, little empirical research or tangible evidence exists of these techniques transfusing from public to private sector. Therefore, although policies do exist to shape the strategic form and direction of the industry, at a tactical level, with particular regard to BIM, they must be deemed inadequate as to its ability to collaborate, provide and retrieve accurate information in the model

The chapter reviews the background of the construction industry. This is to provide a clear picture as to the current status and policies of the industry. Subsequently, it is able to link how Building Information Modelling (BIM) can be fit and applied into the local scenario.

The increased demand and interest in using BIM is a global trend in architectural, engineering, construction and operation (maintenance) sectors (AECO), which is contributed by the enhancement and development for technical and non-technical aspects of BIM. BIM has been described as a modelling tool where different dimensions, information and data can be added into it as deemed fit. Many different parties are required to collaborate and provide information and data available to form the model.

Malaysia, a constitutional monarchy and parliamentary democracy, has a population of 29,567,951 (Malaysia-Gov, 2013) with an annual GDP per capita of US$10,388 (Ministry-of-Finance-Malaysia, 2012). The country became an independent nation in 1957 comprising the former Federated Malay States, the British colonies of Penang, Malacca and Singapore as well as the former British-ruled territories of Sabah and Sarawak on the island of Borneo. Although rich in history, culture and civilisation prior to British involvement (18th Century - 1957), British influence is obvious to Malaysia, and notably so in construction, which the Malaysian Government describes as being closely modelled on the UK's system:

John Rogers, Heap-Yih Chong, Christopher Preece, Chai Chai Lim and Himal Suranga Jayasena

"in that its industry structure, systems, practices and procedures, remain as those which were introduced by Great Britain. The project procurement and administrative arrangements in use have also been inherited from the United Kingdom (UK)" (Malaysia-Gov, 2009, p. 2).

Many of the laws and legislations from the colonial era, and several appear to be not suitable to the development of the country, and hinder the usage of appropriate materials and technologies (Ofori, 2000).

Initially dependent on exports of raw materials, Malaysia has successfully diversified its economy into manufacturing, services, and tourism. Further efforts are being made to transform the country into a 'Knowledge-Economy' by moving up the value chain through knowledge and innovation (The-Economic-Planning-Unit, 2010). Table **1** below describes the sizes of Malaysia's economic sectors relative to each other and to global averages.

Table 1: GDP composition by sector

GDP composition by sector	Malaysia	agriculture: 9.1% industry: 41.6% services: 49.3%	2010 est.
	World Average	agriculture: 5.7% industry: 30.7% services: 63.6%	
the percentage contribution of agriculture, industry and services to total GDP			

Source: (U.S.A-Gov, 2011)

Malaysia, is the world's 28th largest producer of oil (U.S.A-Gov, 2011). In 2012, Petronas, the national oil producer, exported significant quantities of oil and gas and was the major source of tax revenues to the national treasury. Some of this revenue is spent on domestic energy subsidies, and Malaysians enjoy some of the lowest energy prices in the world. Similarly, Malaysia also subsidises water and several essential consumer and food items. Government policies are now aimed at gradually eliminating the subsidies for the transition to a sustainable cost model by 2020 (The-Economic-Planning-Unit, 2010).

Malaysia has a clearly stated objective of becoming a developed nation by 2020 with a Gross National Income per capita of US$15,000. National development is driven by a five year national plan. The 10th Malaysia Plan 2011-2015 (The-Economic-Planning-Unit, 2010) continues to stress the importance of infrastructure development. There is a strong emphasis on sustainable development as demonstrated by the introduction of tax incentives for boosting the uptake of the Green Building Index certification and the provision of sustainable cities.

As part of the strategic approaches to becoming a developed nation by 2020, the Malaysian Government unveiled the Economic Transformation Programme (Malaysia-Government, 2011) which has twelve subservient National Key Economic Areas (NKEA). Several of the eleven industrial sector NKEAs require supportive infrastructure or property development to achieve their objectives. The twelfth NKEA relates to the increased physical development of the Greater Kuala Lumpur/Klang Valley (Malaysia-Government, 2011).

In line with global trends, Malaysia witnesses a rapid growth of urbanization. As shown in Table **2**, urban dwellers accounted for 72% of the total population in 2010. It continues to grow at a rate of 2.4% annually (U.S.A-Gov, 2011). This is consistent with Malaysia experiencing increasing urbanisation:

Table 2: Scale and rate of urbanisation

Urbanisation	Malaysia	Urban population: 72% of total population (2010) Rate of urbanization: 2.4% annual rate of change (2010-15 est.)	2010
	World Average	Urban population: 50.5% of total population (2010) Rate of urbanization: 1.85% annual rate of change (2010-15 est.)	2010
Urban population, describes the percentage of the total population living in urban areas. The rate of urbanization describes the projected average rate of change of the size of the urban population over the given period of time.			

Source: (U.S.A-Gov, 2011)

2.1. INDUSTRY ENVIRONMENT DESCRIPTION

The construction industry contributes significantly to the Malaysian economy. The industry creates economic wealth and enhances the quality of life of the population through the provision of economic and social infrastructure (CIDB, 2008). The industry provides some eight hundred thousand jobs, acts as an economic catalyst and generates a multiplier effect to allied sectors and services (CIDB, 2006).

Chan (CIB-W055, 2010, p. 108) describes *"The output multiplier for building and construction is ranked highest amongst all eleven sectors of the Malaysian economy"*, and *"figures indicate that each RM 1.00 of total output created RM 0.32 of value-add"*. Chan concludes, *"no surprise that the government has decided to invest in infrastructure as part of the economic stimulus packages in response to the 2008 global financial crisis"*.

However, as shown in Table **3** below, by comparison to the global average for Gross Fixed Investment, Malaysia is comparatively low:

Table 3: Gross fixed investment

Gross Fixed Investment	Malaysia	20.1% of GDP (2010 est)
	World Average	23.4% of GDP (2010 est.)

Source: (U.S.A-Gov 2011)

Table **4** below itemises the number and spend on projects in specified sectors:

Table 4: 2011 project sector allocation

Year 2011 Projects									
No.	**Value (RM, mil)**	**Project Category**							
		Residential		**Non-Residential**		**Social Amenities**		**Infrastructure**	
		No.	**RM**	**No.**	**RM**	**No.**	**RM**	**No.**	**RM**
5,555	77,270	1,567	18,721	1,931	29,022	616	4,902	1,441	24,625

Category	Definition
Residential	Residential properties including developments with shop houses
Non-Residential	Commercial and industrial properties, including agricultural and retail
Social Amenities	Civic and social facilities, including education, health, community centres.
Infrastructure	Civil and civic infrastructure; utilities, transport networks and nodes, flood prevention, national projects

Source: (CIDB, 2012a)

Table **5** provides granularity as to the distribution of project sizes. It shows that 72% of projects are clustered in the middle financial value range of RM1-300 million.

Table 5: 2011 project financial size distribution

Type	**Value RM Mil**	**Project Financial Size**								
		0.5-1	**1-5**	**5-10**	**10-50**	**50-100**	**100-300**	**300-500**	**500-1,000**	**1,000+**
Residential	18,721	208	1,633	1,602	8,400	3,211	2,852	813		
Non-Residential	29,022	265	2,274	1,823	6,783	3,316	3,332	1,189	1,382	8,658
Social Amenities	4,902	87	739	515	2,646	482	117	318		
Infrastructure	24,625	162	1,507	2,139	6,501	2,759	2,922	338	2,159	6,139
Total	77,270	723	6,153	6,080	24,328	9,768	9,222	2,659	3,541	14,797

Source: (CIDB, 2012a)

Table **6** below provides statistics on the number and value of work undertaken by procurement type, and by client type. Of note is that in 2011, by financial value, 95% of all property sector work in Malaysia was carried out *via* the conventional, or traditional, procurement route. Alternative routes are comparatively negligible. Private, conventionally procured projects account for 61% of all works in Malaysia.

Table 6: 2011 number and value of projects awarded by client sector and procurement type

Type	**No.**	**Value (RM, mil)**	**Public Project**		**Private Project**	
			No.	**RM**	**No.**	**RM**
Conventional	5,408	73,188	1,460	16,566	3,948	56,622
Design & Build	100	3,421	51	1,922	49	1,499
Turnkey	39	571	9	168	30	404
Build, Operate & Transfer	2	14	-	-	2	14
Engineering, Procurement, Construction & Commissioning	6	76	-	-	6	76
Total	5,555	77,270	1,520	18,656	4,035	58,615

Source: (CIDB, 2012a)

Table 7 decomposes projects into typologies. New projects account for 90%, by financial measure, of all projects undertaken. New projects delivered through conventional procurement accounts for 85% of all work.

Table 7: 2011 number and value of projects awarded by type of work and type of contract

Procurement Type	No.	RM Mil	Type of Work					
			New	Upgrade	Expansion	Maintenance	Repair	Renovation
			RM	RM	RM	RM	RM	RM
Conventional	5,408	73,188	65,843	3,825	1,208	726	823	764
Design & Build	100	3,421	3,003	176	171	14	32	25
Turnkey	39	571	437	8	0.76	4.6	118	2.95
Build, Operate & Transfer	2	14	13.96	-	-	-	-	-
Engineering, Procurement, Construction & Commissioning	6	76	75.94	-	-	-	-	-
Total	5,555	77,270	69,373	4,008	1,379	745	973	791

Source: (CIDB, 2012a)

Thus, a typical project in Malaysia can be considered as a new build project, costing in the region of RM50 million, delivered through the conventional form of procurement for a private client.

2.1.1. Current Construction Industry Status

The Ministry of Finance's Third Quarter 2012 (Ministry-of-Finance-Malaysia, 2012, p. 8) report provides the following account:

"Construction activity remained robust. The construction sector continued to record a strong growth of 18.3% in the third quarter of 2012 (Q2 2012: 22.2%) contributed by the robust expansion in the civil engineering and residential subsectors. The civil engineering subsector grew at a double-digit pace of 28.5% (Q2 2012: 39.8%) amid accelerating works in large-scale infrastructure projects, namely the Sungai Buloh - Kajang Line MY Rapid Transit, LNG Regasification Terminal, Melaka and Manjung coal-fired plant. The residential subsector also grew strongly by 16.4% (Q2 2012: 20.1%) mainly supported by stronger high-end housing starts in Klang Valley, Sabah and Penang. Meanwhile, the special trade subsector grew 8.6% (Q2

2012: 11.4%), while the non- residential subsector rose 13.4% (Q2 2012: 11.7%) due to increasing construction of commercial and industrial buildings, particularly the Polycrystalline Silicon factory in Sarawak".

Set against this however is the CIDB's Building Materials Cost Index (CIDB, 2011) that shows 2011 prices still several percent below those of 2008.

Any macro-economic data emanating from the built environment must be treated with caution. Although the United Nations provides a framework for consistent international metrics and national accounting (United-Nations, 2001) the reality of national accounting means that interpretation and application vary considerably from country to country. As an example, Malaysia routinely declares construction's GDP contribution as between 2.7 - 3.9% (Ministry-of-Finance-Malaysia, 2011) shown in Fig. **1**. As identified in Section 2.4.3 above, the traditional global norm is between 5 - 10%. Professor Ofori of the Faculty of the Built Environment, National University of Singapore, considers this an accounting anomaly (Ofori, 2011).

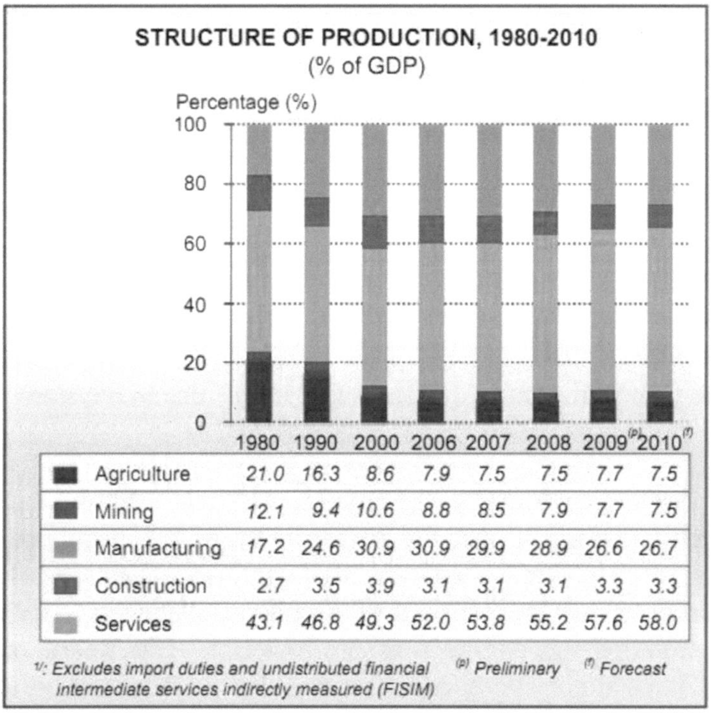

	1980	1990	2000	2006	2007	2008	2009[p]	2010[f]
Agriculture	21.0	16.3	8.6	7.9	7.5	7.5	7.7	7.5
Mining	12.1	9.4	10.6	8.8	8.5	7.9	7.7	7.5
Manufacturing	17.2	24.6	30.9	30.9	29.9	28.9	26.6	26.7
Construction	2.7	3.5	3.9	3.1	3.1	3.1	3.3	3.3
Services	43.1	46.8	49.3	52.0	53.8	55.2	57.6	58.0

Figure 1: Malaysian GDP statistics 1980-2010. (Source: Malaysia-Gov, 2011).

2.1.2. National Development Maturity

The nature of the built environment and the construction industry change as the nation develops (Ruddock, 2007). For lesser-developed nations, the GDP share of construction increases as the nation develops, but eventually the contribution diminishes as the country meta-morphs through industrialisation and into a service based economy Lopes, J. in CIB-W055 (2010, p. 151).

> *'In low income countries, construction output is low. As industrialisation proceeds, factories, offices, infrastructure and houses are required, and construction as a percentage of gross domestic product reaches a peak in middle income countries. It then tapers off as the infrastructure becomes more developed and housing shortages are less severe or are eliminated'* (Ruddock, 2007, p. 9).

There are inherent difficulties in the metamorphosis that developing countries experience (Ofori, 1993b), especially, economic stress, resource deficiencies, and institutional and legal weaknesses (Ofori and Lewis, 1993). Ofori (2007) reiterated and extended the negative characteristics that hinder progress and overall project performance.

Mohamed and Egbu (CIB-W102, 2010, p. 256) reported that, *"Based on a critical review of literature on knowledge sharing in government agencies, dysfunctional bureaucracies, training, utilisation of information technology and adaptation to culture were identified as key issues"*. By extension, this finding suggests institutional limitations and capability constraints on the optimal exchange of knowledge between all stakeholder groups within the built environment delivery process.

2.1.3. The Malaysian Construction Industry Structure

There are three principal statutory bodies governing the Malaysian construction sector:

The Board of Architects Malaysia (Lembaga Arkitek Malaysia - LAM) is a statutory authority responsible for the enforcement of the Architects Act 1967. LAM accredits, registers and regulates the conduct of entities that perform architectural works or services. (Board-of-Architects-Malaysia, 2012). Providing a parallel function for the engineering profession is The Board of Engineers Malaysia (BEM); a statutory body constituted under the Registration of Engineers Act 1967 (Board-of-Engineers-Malaysia, 2012). The third body is the Construction Industry Development Board more fully described in Section 1.3.3.

2.2. CONTEMPORARY REVIEWS OF MALAYSIAN CONSTRUCTION SECTOR

Razak *et al.,* (2010, p. 299) state *"very little research has been carried out by academics and practitioners on the problems faced by the local[Malaysian]construction industry"*. Adding, *"Construction industry players are not conducting Research and Development (R&D) and in Malaysia most of the R&D is carried out in academic institutions. But areas covered are usually not in accordance with industry needs"*.

Of the little research accomplished, Chan (2009, p. 1232) notes that it is predominantly project performance and construction company centric and notes, *"Other stakeholders, such as clients, suppliers, regulatory authorities and the community were not assessed or taken into account"*.

Riazi, Skitmore *et al.,* (2011) cite the CIDB as describing the construction industry as underachieving, citing delay rates of up to 80% on public projects as a significant symptom. Razak *et al.,* (2010) cite, *"Abdul Rahman et al., (2006), have found 45.9 percent delays in the completion dates during the construction stage"*. Abdul-Rahman *et al.,* (2006, p. 130) identifies financial issues as the primary source of delay. As a contributory and possibly causative factor, they identify lack of human resource capability and capacity at all levels of the industry. Abdul-Rahman *et al.,* (2006, p. 125), notes, *"the construction industry regularly shows lower levels of productivity when compared with the manufacturing industry"*. This is supported by Chan (2009, p. 1243) who utilises Kaplan and Norton's strategy map to provide an industry level analysis of the Malaysian construction sector, and summarises thus:

> *"Data for the 2006 base year indicate low annual increases in productivity, inadequate safety performance, low investment in research and development, and a disappointingly low number of construction companies certified to quality, environmental and occupational health standards. Nearly half the workforce is unskilled"*.

Riazi, Skitmore *et al.,* (2011) and Taib (2010) consider the primary causal problem to be the practices and management systems of the industry and conclude that the industry is adversarial, adopting a 'blame game' posture and that, *"This suggests the need for a transformation in the way projects are managed"*.

A comprehensive assessment of construction labour productivity in Malaysia by Chia *et al.,* (2010) analysed periodic government survey and census data and reported:

> *"The added value per employee of the construction sector has increased between 1998 and 2004 and decreased between 2004 and 2007. The total output per employee follows the trend of the value-added per employee, i.e. it rises between 1998 and 2004 and falls between 2004 and 2007. However the value added per employee is in contraction. The capital intensity increased, but the capital productivity did not catch up. The added value content of contracts is 29.81 in 2007, compared with 38.51 in 1997"* (CIB-W055, 2010, p. 206).

The same authors noted in CIB-W055 (2010, p. 267):

> *"Malaysia ranked 50th and 63rd position among the 79 selected economies on real and nominal measurement respectively. It generally performed satisfactory, however, when compared with ASEAN countries and other similar economies. However, it has only 49 percent of the average nominal productivity level of upper-middle income economies and is marginally below the average real productivity level achieved by upper-middle income economies. In addition, the ranking comparison also shows that Malaysia has invested relatively higher in construction than the productivity it achieved".*

Regarding the financial health of the commercial construction companies within the sector, Ab Halim *et al.*, (2011) quantitative research deploys financial ratio analysis and identifies that most companies are in a parlous state:

> *"Overall results reveal the extreme financial difficulties these companies may face to sustain their business in the coming years. Majority of the construction companies do not have sufficient cash capital to finance their construction works, companies benefit low profit margin from construction projects, and companies were highly dependent on debt capital to finance their construction costs"* (Ab Halim *et al.*, 2011, p. 200).

2.3. SECTOR IMPROVEMENT INITIATIVES

2.3.1. Malaysian Lean Construction

Lim's Masters' thesis (2008) explores the appreciation and application of lean construction principles in the Malaysian construction industry. Although noting that very few direct implementations had occurred, there was a small yet enthusiastic appreciation of the potential benefits that lean could bring.

Furthermore, Lim contends that aspects of lean were being implemented and were known by different terminology or were business-as-usual practices, such as waste reduction. However, Lim identifies that considerable barriers existed to greater implementation, requiring changes in mind-set, a re-emphasis on client needs and a re-alignment of the industry to become client centric. The peninsular wide questionnaire survey was limited to contractors' staff and received only forty completed responses. No information was provided regarding project size, type or procurement system.

A further quantitative study that sought to identify barriers to lean construction in Malaysia was undertaken by Abdullah *et al.,* (2009). Limited to design and project management teams on the east coast of the peninsular, the responses of ninety three individuals identified that lack of understanding and lack of senior management buy-in were the principal barriers to greater adoption.

Notwithstanding the previous research, Ibrahim *et al.,* (2010a, p. 235) continued to advocate lean construction and the 'last planner system' in Malaysia, stating, "*LPS is simply without equal and can be applied equally in any industry around the globe*". This however is a concept paper that provides an overview of the Malaysian industry as opposed to empirical research.

2.3.2. Malaysia's Industrialised Building Systems

Industrialised Building Systems have a history in Malaysia dating back to the 1960s (Fathi *et al.*, 2012a). The IBS market is a developing and important market in Malaysia particularly since all public-sector projects must attain no less than 70% IBS content under the Treasury Circular SPP 07/2008. This policy was created to build a momentum and to establish demand for IBS components, thus reducing costs through economies of scale (Abdullah and Egbu, 2010). The IBS market in Malaysia is estimated to increase to RM 85 billion by 2015, yet acknowledging its potential, the construction industry is still not rapidly embracing IBS (Construction-Research-Institute-Malaysia, 2011).

In 1999, Malaysia's Construction Industry Development Board (CIDB) released the IBS Strategic Plan 1999 that evolved into the IBS Roadmap 2003-2010 (CIDB, 2009), with the intent of increasing the use of the Industrial Building System (IBS) which deploys *"techniques, products, components, or building systems which involve prefabricated components and on-site installation"* (CIDB, 2005). Chia, Skitmore, Runeson and Bridge CIB-W055 (2010, p. 210) describe characteristics as:

> *"encouraging the use of alternative construction materials and technology under the IBS and designs based on the modular coordination concept in housing construction; using IBS components in the construction of affordable homes and in Government building projects and enforcing the use of modular coordination concept through Uniformed Building By Laws by the local authorities, increasing usage of IBS components in Government building projects from 30 percent to 50 percent commencing 2005 and giving accelerated capital allowance for capital expenditure on moulds to manufacture IBS components".*

The importance of IBS is highlighted under the Construction Industry Master Plan 2006-2015 (CIMP, 2006-2015) Strategic Thrust 5: Innovate through R&D to adopt new construction methods.

The IBS strategy is also designed to implement social policy (Fathi *et al.,* 2012a). In 2007, 69% of Malaysia's construction workers are foreign nationals and:

> *"the main reason for Malaysia to move into industrialised construction is due to the influx of foreign labour doing manual jobs in construction.... It is hoping the industrialisation of the industry through mechanisation, pre-fabrication and automation will reduce the number of foreign labour and it eventually will be replaced by high skilled local workforce"* (CIB-TG57, 2010, p. 65).

2.3.3. Malaysia's IPD/Relational Approaches

Contract is important in balancing risks involved between the contracting parties (Harris *et al.*, 2006). Different project characteristics will have different contracts, some of the factors are the length of the project, its complexity, its size and that the price agreed and the amount of work done may change as it proceeds, it is also particular to each country (Robinson *et al.*, 1988).

The academic literature distinguishes between two types of collaborative agreements: contractual and relational (Wang, 2008, Kent and Becerik-Gerber, 2010, CIB-W092, 2010). Contractual elements provide guarantees and form the basis for most standard forms of contract used throughout the construction industry. Relational factors bring trust and tend to favour learning among actors (CIB-W113, 2010).

The contractual element seeks to pre-specify solutions to as many eventualities as possible to reduce unforeseen disputes and claims (Chong and Zin, 2012). However, this generates a disparate and adversarial industry, where clear and rigid rules restrict interaction between parties (Latham, 1994). This does not help build a collaborative and long term relationship between the contracting parties as the contracts focus on self-interest only (Egan, 1998). The inadequacies of traditional contracts have resulted in the emergence of relational contracts that aims to achieve synergy between self-interest, collaboration and fairness (Bongiorni and Cohort, 2011).

Integrated Project Delivery (IPD) has emerged as an applied, relational contract delivery system, (Chan *et al.*, 1999, Matthews and Howell, 2005) that demonstrates success on real-life projects, (Post, 2011), through embracing maximum relational contracting techniques. As a procurement system it seeks to align interests, practices, and goals through a team-based approach. The primary team members generally include the architect, engineers, the main contractor and key subcontractors.

Matthews and Howell (2005, p. 61) describe it as:

"a relational contracting approach that aligns project objectives with the interests of key participants.a clever solution to the tough organizational and contracting problems faced in today's market. It relies on careful participant selection, transparency and continuing dialog. Construction consumers might consider rethinking their contracting strategies to share more fully in the benefits".

As a system, it is still in its infancy, yet its leading advocates, the Construction Managers Association of America (Thomsen and FAIA, 2010), ascribe relational improvements beyond Design and Build, in which system the contractor is predominant, to a fully collaborative, mutually supportive, egalitarian system that derives additional benefit from the whole team.

Acknowledging the UK's pre-1957 colonial legacy to Malaysia's construction industry, the Malaysian Government recognises that the construction industry is resultantly adversarial by nature and has a fragmented and generally discordant supply chain (Malaysia-Gov, 2009), as such, initiatives have been attempted to integrate the supply chain, with relational contracting and partnering efforts being encouraged. As can be seen from Table **6**, in 2011 89% of publicly procured works was *via* conventional procurement with the remaining works procured through either design and build or turnkey contracting.

Ismail *et al.,* (2012) identify that specimen examples of relational contracting do exist in the private sector. They also remind that iconic Malaysian mega-projects, such as Petronas Twin Towers, the SMART Tunnel and Kuala Lumpur International Airport, were delivered through relational contracts, signifying that relational contracting can be appropriate and suitable for Malaysian cultural practices, as expanded by Faisol (2010). However, the evidence suggests that relational contracting has yet to find traction within the greater Malaysian construction industry.

Zuhairah *et al.*, provide a précis of the legal context for Malaysian construction projects:

> *"There are various standard forms of construction contract, which the parties can refer, adopt or incorporate in Malaysia three main standard forms contract, which are referred to in Malaysia; the PWD 203A (2007 edition), the PAM 2006 and the CIDB Standard Form of Contract for Building Works 2000. For the public sector, the main reference is normally the PWD 203 form, whilst for the private sector; the PAM standard form contract and the CIDB Building Works contract"* (CIB-W113, 2010, p. 50).

In their standard form, these contracts are intended for conventional procurement. At the time of writing, no standard forms were appropriate for relational procurement routes.

2.3.4. Malaysian Hybrid Systems

The 8th IBS Roundtable, featuring leading construction experts from academia and CIDB (CREAM), (Kamar, 2011a) set as its 2011 theme: Mechanisation through Building Information Modelling. It identified BIM as a critical enabler in the pursuit of IBS, and noted the symbiotic relationship between BIM and IBS, although IBS remains the dominant sector improvement strategy (Kamar, 2011b).

The CIDB's research arm, CREAM, have subsumed BIM into the wider construct of Integrated Design Delivery Solution (IDDS) which also contains IPD (CREAM, 2012), and provides an overview of some proposed implementations of BIM and IPD in Malaysia (CIDB and CREAM, 2011), however, at the time of writing, no published literature was available outside of the project descriptions (CREAM, 2012).

2.4. MALAYSIAN GOVERNMENT CONSTRUCTION INDUSTRY POLICIES

Having reviewed the Malaysian construction industry, positioned the industry within the national context and described the character and status of the industry, this section reviews contemporary issues within and around the sector.

2.4.1. Environmental Sustainability

The Malaysian Government has attempted to enshrine the principles of sustainable development into national policy plans. The recent national development plan for 2011-2015 (The-Economic-Planning-Unit, 2010) has a strong emphasis on sustainable development as demonstrated by the introduction of tax incentives in order to boost the uptake of Green Building Index certification and the provision of sustainable cities (REHDA, 2011). This is supported by the Economic Transformation Programme (Malaysia-Government, 2011) which is tasked with achieving developed nation status by the year 2020.

Despite the positive policy rhetoric, there is no evidence of legislation ensuring mandatory practices of sustainability for the building sector. Sustainability efforts in the design, construction and operation of buildings are currently carried out on a voluntary basis.

In Malaysia, the most applied sustainability assessment tool for buildings is the Green Building Index (GBI). Designed by the Malaysian Institute of Architects (PAM) and the Association of Consulting Engineers Malaysia (ACEM), GBI fits Malaysia's current social, infrastructure and economic development (Green-Building-Index, 2012). It includes tools to support assessment of buildings from single residential to industrial and township level and some supportive design guides. Table **8** shows the preffered GBI tool in Malaysia.

Table 8: Building sustainability accreditation overview

Sustainability Assessment Tool	No. of Certified Buildings	Source *
Green Mark	9	http://www.greenmark.sg
GBI	55	http://www.greenbuildingindex.org
LEED	13 Certified 26 Registered	http://www.usgbc.org

*Databases visited on 14 January 2013

Since the introduction of GBI in 2009, uptake remains low (Nizarudin *et al.*, 2010) and sustainable design is still perceived as an added cost to a project without any obvious commercial benefits (Abidin, 2010b).

Cost is one of the most critical factors of property developers' decision making process (Abidin, 2009, Abidin and Jaapar, 2009, Abidin, 2010a). GBI estimate a cost increase of three to five percent depending on the level of certification to be achieved (Nizarudin *et al.*, 2010). According to World Green Building Council (Managan, 2011), green building practices can reduce a building's operating costs by as much as 9 per cent, increase a building's value by 7.5 per cent and realise a 6.6 percent increase in return of investment. Against this, Nielsen's 2011 Global Online Environment and Sustainability Survey placed Malaysian consumers as the second least likely group among their ASEAN counterparts to willingly pay more for eco-friendly products (Nielsen, 2011). Therefore it is likely that any improvement will be driven by the Government:

> *"The commitment from all stakeholders can transform the Malaysian construction industry into one that is not a threat to the environment, but meets the human need for development in harmony with the nature. Development will continue to expand to meet socio-economic obligations, but the focus should be on the quality of the development that is pursued"* (Malaysia-Gov, 2007).

2.4.2. Value Management

In 2009 the Malaysian Government issued the "Value Management Guideline Circular 3/2009" of which, Saifulnizam (2010, p. 391) states:

> *"[it] mandates that for every [public] construction project valued at more than RM 50 million, a VM study is required to be undertaken. This marks the beginning of a new paradigm for VM in the Malaysian construction industry, which is expected to allow for greater use and development of VM in the sector".*

Value Management (VM) is a method for enhancing the value in projects, products, processes and systems (Male *et al.*, 2005).VM is a multi-disciplinary effort (Kelly and Male, 1993) from the whole life cycle perspective. The value will made explicit and be assessed with a rating guideline determined by the employer (Kelly *et al.*, 2004). Che Mat (Che Mat, 2010, p. 1) cites Dell I'sola, *"value is defined as the most cost effective way to reliably accomplish a function that will meet the user's needs, desires and expectations".*

VM was first introduced in Malaysia in 1986 (Che Mat, 2010). Since its introduction, and despite some gains can be generated from this method, it is yet to be widely practiced in the country (Jaapar *et al.*, 2009, Tek, 2004). They conclude that a knowledge gap exists between the current practice of VM in the Malaysian construction industry in comparison to practices employed in developed countries. They suggest that this is largely due to the lack of understanding and consistent implementation of VM, compounded by a resistance to change by the parties involved during VM workshops. Hamid *et al.*, (2011, p. 421) describe some of the shortcomings of VM in Malaysia: *"VM is criticized for the fact that it focused solely on value for money, lack of creativity and fails to include the concept of risk management at the early project stage".*

Some client organizations in Malaysia have been applying certain aspects and concepts of the VM methodology in their project operations and many construction industry professionals have acknowledged that VM contributes to the achievement of value for money (Jaapar *et al.*, 2009). It will be a positive indication of practice for VM in the Malaysian construction industry (Jaapar *et al.*, 2008).

Chapter 3: Sri Lankan Construction Industry

CHAPTER 3

Sri Lankan Construction Industry

Abstract: Sri Lanka is a republic and a unitary state governed by a presidential system. The colonial influence, primarily of British, is still notable in the legal, educational and industrial sectors in Sri Lanka. Procurement systems adopted in Sri Lankan Construction Industry are similar to those in the United Kingdom. Pricing efficiency in the Sri Lankan construction industry is significantly low. The Institute for Construction Training and Development (ICTAD) is the prominent government institute in Construction Industry. Nevertheless, it is obvious that the industry requires more structured and effective institutional and regulatory framework. Besides, the Sri Lankan construction industry will require embracing the contemporary development in knowledge and technology.

This chapter describes the background and current development of Sri Lankan Construction Industry. As Sri Lanka is a much less-developed country, limited data are available for discussions. Yet, this will provide some fundamental references for subsequent discussion on BIM case study.

3.1. THE DEVELOPMENT OF SRI LANKAN CONSTRUCTION INDUSTRY

Sri Lanka, officially the Democratic Socialist Republic of Sri Lanka, is a republic and a unitary state governed by a presidential system. It is an island in the north Indian Ocean close to south Indian coast and Maldivian islands. The island with land area of 65,610 (62,705 excluding inland waters) square kilometres had a reported population of 20.5 million in year 2013 with a 0.8% population growth at the time. Being a developing country, Sri Lanka has a significantly high literacy rate of more than 95% and average life expectancy of 75 years (2012 data). About 12% of the population is 60 years or above and a quarter of the population is below 15 years (Central Bank of Sri Lanka, 2014).

The documented history of Sri Lanka dates back over 3,000 years (van Horen and Pinnawala, 2006, p. 309). "*There is secure evidence of settlements in Sri Lanka by 130,000 years ago, probably by 300,000 BP and possibly by 500,000 BP or earlier*" (Deraniyagala, 1996, p. 277). "*Its civilization has been shaped largely by that of the Indian subcontinent. The island's two major ethnic groups, the Sinhalese and the Tamils, and its two dominant religions, Buddhism and Hinduism, made their way to the island from India, and Indian influence pervaded*

John Rogers, Heap-Yih Chong, Christopher Preece, Chai Chai Lim and Himal Suranga Jayasena

such diverse fields as art, architecture, literature, music, medicine, and astronomy" (Arasaratnam, 2014).

"Sri Lanka was under colonial rule, variously from the Portuguese, Dutch, and British from 1505 until 1948. Sri Lanka (then known as Ceylon) achieved independence from the British in 1948" (van Horen and Pinnawala, 2006, p. 310). The colonial influence, primarily that of British, is still notable in the legal, educational and industrial sectors in Sri Lanka.

Procurement systems adopted in Sri Lankan Construction Industry are similar to those in the United Kingdom. However, collaborative systems such as Partnering are not found in Sri Lanka. Conventional (Traditional) Method remains the dominant method since colonial era. Design and Build (D&B) is the next alternative and it has been used in 20-30% of building projects and there has not been an increasing trend detected. Major impediments for the development of D&B were the lack of awareness of alternatives and lack of promotion of the method by the government (Joseph and Jayasena, 2008).

Pricing efficiency in the Sri Lankan construction industry is significantly low. Tan and Suranga (2008) found evidence of the existence of large winner's curses in the Sri Lankan construction industry. Standard deviation of bids distribution was 16% of the average bid, suggesting that winning bids were often 16% or more below the average bid. While it is generally thought that larger winner curses could cause adverse claim attitudes, Jayasena and Uhanowitage (2008) showed that this effect is minimal or non-existent in Sri Lankan industry.

Construction industry employs about 7-8% of the labour force of the country. There are no strong labour unions to represent skilled and semi-skilled labourers. However, there are professional institutes to represent all key AEC professions. The Institute for Construction Training and Development (ICTAD) is the prominent government institute in Construction Industry. Its mission is "*to ensure dynamic, professional, and reliable value added services to the nation, through regulation and facilitation of the development of construction industry resources and promotion of quality standards, to meet local and global requirements for sustainable national development*" (ICTAD, 2014). The diffused nature of the construction industry has made it necessary for many other government institutes to play important roles in the industry both as clients and regulators.

The construction industry comprises of adequate institutional infrastructure. However, it is often questioned if the institutional and regulatory framework is appropriate. Some professionals argue that "*seemingly disparate authorities control the sector without efficient coordination among them which leads to delay, wastage and sometimes failure. The standards and specifications in the local construction industry have evolved into a near comparable level with contemporary international standards. However, the use of some standard documents such as the Standard Method of Measurement (SMM & CESMM) and the standard preliminaries bill are rare in the local industry*" (2014, p. 2).

ICTAD has published a range of standard documents to be used in local construction industry. The ICTAD Standard Bidding Documents (SBDs) are developed by referring and reviewing international standard documents. The Conditions of Contract are a part of SBD. Most of the documents closely resemble their FIDIC (International Federation of Consulting Engineers) counterpart and therefore offer ease of adoption.

Sustainable construction is promoted by the Green Building Council of Sri Lanka (GBCSL). "The aim of GBCSL is to transform the Sri Lankan construction industry with green building practices and to fully adopt sustainability as the means by which the environment thrives, economy prospers and society grows to ensure the future wellbeing of the country. The Green Environmental Rating System for Sri Lanka has been formulated as a 'home-grown system' with all norms acceptable to leading rating systems" (GBCSL, 2014).

It is notable that "the complexity of the construction industry and its projects has dramatically increased in terms of products and processes while there is also an increase in terms of magnitude. The increased awareness of and market pressure on sustainable development, the self-consciousness of adverse impacts from the failure to reach sustainability goals, and the skyrocketing energy costs have led to clients demanding whole life value from their projects" (Daily News2014). To meet the demands, the Sri Lankan construction industry will require embracing the contemporary development in knowledge and technology.

3.2. AN OVERVIEW

The construction industry mainly consisted of building constructions, highway constructions, bridge constructions, services supply and drainage systems, dredging, reclamations, and so on. Most of the construction activities generally can be categorized under civil engineering. After the war in 2009, Sri Lanka has progressed very well in the construction industry, particularly in the building construction where a great demand for property development. ICTAD plays an important role and formulated many industry standards to drive the construction industry forward.

Overall, the industry is on the right track. Many megaprojects have been initiated with collaborations or partnerships with neighbouring countries, namely, China and India. The government has also promoted the use of environmentally friendly and cost effective technologies and building system in the country.

CHAPTER 4

BIM - Building Information Modelling

Abstract: Building Information Modelling (BIM) is an abstract concept. The preferred definition of BIM is "Building Information Modelling is digital representation of physical and functional characteristics of a facility creating a shared knowledge resource for information about it forming a reliable basis for decisions during its life cycle, from earliest conception to demolition" (Snook, 2011). Different dimensions and capabilities of BIM are articulated, such as 4D, 5D, 6D, *etc*. Interoperability issues have been discussed with Industry Foundation Classes (IFC) specification and Construction Operations Building Information Exchange (COBie). BIM maturity models have been compared. Subsequently, the chapter reviews a brief historical perspective on BIM software. It then proceeds to describe current and commonly available BIM software. Eventually, it concludes with an overview of training programmes that are pertinent to the obtaining of BIM software skills.

This chapter introduces BIM's philosophies and current developments. It reviews different dimensions and capabilities of BIM with certain examples for reference. Subsequently, it also discusses other contemporary issues related to BIM's adoption, practice and software.

4.1. DEFINITIONS OF BIM

BIM is a rapidly emerging phenomenon that may greatly influence the evolution of the construction industry. Accordingly, this Section starts with a review of existing definitions of BIM and proceeds to select one definition for the purposes of this book. It then proceeds to define the characteristics of BIM.

As observed by the USA's National Institute of Building Sciences (National-Institute-Building-Sciences, 2007, p. 1), *"There are currently almost as many definitions for BIM as there are people implementing them"*.

Following Neuman (2006, p. 54), BIM is an abstract concept cluster, it is understood as a web of meaning comprised of many other concepts. It is commonly used as a generic term encompassing a wide and expanding cluster of concepts. The following provides a range of definitions.

The NBS (The-NBS, 2013a, p. 1) seek to define BIM by listing some of its functional outputs. Some potential is alluded to and an indication of the broad

John Rogers, Heap-Yih Chong, Christopher Preece, Chai Chai Lim and Himal Suranga Jayasena

categories of user or disciplines that can draw benefit from its output are identified:

> *"Building information modelling covers geometry, spatial relationships, light analysis, geographic information, quantities and properties of building components. BIM data can be used to illustrate the entire building life cycle, from cradle to cradle; quantities and properties of materials can be extracted easily and the scope of works can be easily defined. Furthermore systems, assemblies and sequences can be shown in a relative scale to each other and relative to the entire project".*

Gu and London (2010, p. 988) provide a technology dominant perspective that gestures towards the expected benefits.

> *"Building Information Modelling (BIM) is an IT enabled approach that involves applying and maintaining an integral digital representation of all building information for different phases of the project lifecycle in the form of a data repository. The building information involved in the BIM approach can include both geometric data as well as non-geometric data. BIM is one of the important areas in current Virtual Reality (VR) research and is expected to envision efficient collaboration, improved data integrity, intelligent documentation, distributed access and retrieval of building data and high- quality project outcome through enhanced performance analysis, as well as multidisciplinary planning and coordination".*

The suggested benefits include improved design coordination, reduced design and site conflict, energy and carbon savings, and a valuable information body for facilities and asset management. However, it is a nebulous concept that expands with technical capability and capacity growth. Authorities such as Eastman, (2011), consider BIM as an enabler that may help the building industry to improve its productivity and ultimately value to the Client body (Succar, 2010). buildingSMART (2012b, p. 1) decomposes BIM into three sub-concepts:

- ***"Building Information Modelling**: Is a business process for generating and leveraging building data to design, construct and operate the building during its lifecycle. BIM allows all stakeholders to have access to the same information at the same time through interoperability between technology platforms".*

- ***"Building Information Model**: Is the digital representation of physical and functional characteristics of a facility. As such it serves as a shared knowledge resource for information about a facility, forming a reliable basis for decisions during its life-cycle from inception onwards".*

- ***"Building Information Management**: Is the organization & control of the business process by utilizing the information in the digital prototype to effect the sharing of information over the entire life-cycle of an asset. The benefits include centralised and visual communication, early exploration of options, sustainability, efficient design, integration of disciplines, site control, as built documentation etc. - effectively developing an asset life-cycle process and model from conception to final retirement".*

The UK Government (BIM-Industry-Working-Group, 2011, p. 7) considers the whole-life perspective, extending its purpose to asset management:

> *"In order to improve the measurement and management of public assets, it is recommended that public clients request that specific information be delivered by the supply chain. The specified information set, called COBie, delivers consistent and structured asset information useful to the owner-operator for post-occupancy decision-making".*

Furthermore, the UK Government is committed to making BIM compulsory on all government projects by 2016, believing that adopting BIM systems in the design, construction, operation and refurbishment of buildings will help deliver a low carbon future *"Government as a client can derive significant improvements in cost, value and carbon performance through the use of open shareable asset information"* (UK-Gov, 2011b, p. 15).

However, Brewer (2011, p. 244) identifies that barriers to widespread adoption mean that BIM is infrequently used as the intended panacea and central data repository for all project participants, and that in present practice:

> *"[BIM] may be more feasible for peripheral participants in construction projects, who have self-contained supply chains independent of the project itself".*

Concluding:

> *"the success of BIM in a temporary project organisation (TPO) is dependent upon the presence of participant firms that share compatible technologies, business processes, and cultures, led by people who hold attitudes and display behaviours conducive to collaboration"* (Brewer, 2011, p. 246).

4.2. THE PREFERRED DEFINITION

BIM can be seen from perspectives encompassing the technology, process and human dimensions (Coates *et al.*, 2010). Fundamentally, the building information model is a co-ordinated and consistent data store to which information and knowledge generation procedures - advanced or simple - can be applied depending on specific requirements. The satisfaction of the specific requirements may be contingent upon restructuring of business processes and relationships, and some requirements may as yet be outside of the current capacity and capability of either the technology or human resources, but these are temporary, solvable constraints.

Thus the definition of BIM adopted for this book is one that emphasises the output characteristics, articulates the scope of impact and encourages an expansive appreciation of future potential. It is a definition endorsed by significant bodies including the UK's RIBA, CPIC, buildingSmart and the USA's National BIM Standards Committee (NBIMS) and is universal in its acceptance:

> *"Building Information Modelling is digital representation of physical and functional characteristics of a facility creating a shared knowledge resource for information about it forming a reliable basis for decisions during its life cycle, from earliest conception to demolition"* (Snook, 2011).

4.3. CHARACTERISTICS OF BIM

The above definitions provide an introduction to the 'technology', 'process' and 'people' dimensions of BIM. The following section provides a deeper description of these dimensions.

4.3.1. Technology

At the core of BIM is a repository of data that is systematically organised according to predetermined rules (McGraw-Hill-Construction, 2009). These rules determine the scope, nature, detail and any specified characteristics required (CIB-W078, 2010). As a minimum, key characteristics, known as parameters, of each component or object are recorded, such as; height, length, depth, weight, spatial/geo-spatial location, material/element properties, spatial relationship to adjacent objects (buildingSMART, 2012d). This base data can then be interrogated and manipulated by a CAD application to produce two and three dimensional representations of the building (or infrastructure object) (Eastman *et al.*, 2011). The concept of parametric data about the object is paramount, which is based on a set of rules to determine the performance (Dwyer 2012, p. 36).

Table **9** below provides a summary of BIM capabilities drawn from relevant authorities.

Table 9: Dimensions and capabilities of BIM

Building Information Model Application Areas	**Extended Description**
BIM for Early/Conceptual Design	This can include photo-realistic visualisations to support design process, planning applications and communications with non-expert stakeholders and the general public. These can be virtual walk-throughs and not just static images. Can be extended to include operational user testing, for example, projected customer flows in shopping centres or patient movement in hospitals. Visualisations are commonly used for sales and marketing purposes.
3D and Parametric Modelling	Individual components can be modelled and appraised. These can be linked into their respective systems. For example, a variable air volume (VAV) damper can be individually assessed and then incorporated and assessed as part of the overall HVAC system. The HVAC system can then be incorporated into the entire building model and assessed in conjunction with other services, structural or architectural assets. 3D models provide for the (semi)-automated provision of coordinated services drawings. This enables visual verification of design concepts and can be extended to automatically model human space requirements for access - this is particularly beneficial for maintenance works in plant rooms.
Acoustic and lighting	Interactive, parametric 3D models enable the interrogation and modelling of spaces and surfaces for acoustic and lighting design development.
4D - assembly and operations	Scheduling and sequencing of site activities. This can be extended to include site logistics and temporary works planning. Materials delivery sequencing can be planned, such as site deliveries in congested inner-cities.
5D - estimating and cost	Estimating and cost management. Scenario and Value Management activities can be included.
6D - procurement/supply-chain	Can include e-procurement and integration of the supply chain. This includes the interaction with component and materials suppliers. This can include design information, such as lift or escalator manufacturers, shop fabrication drawings that can be included in the model for 'as-built' record drawings. This

Building Information Model Application Areas	**Extended Description**
	can include information required under COBie described below.
6D - facilities management	Planned and preventative maintenance can be automatically scheduled, complete with requisite information such as supplier details. Can be linked directly to inventory management systems. Reactive maintenance can be aided by the modelling of remedying works complete with all salient supplier information.
6D - operations/facilities usage	Models can be integrated with space usage applications to enable optimised facilities occupation services such as room bookings or business/operational process changes. Real-time monitoring of building usage, including energy usage, in integration with building management systems (BMS).
Structural Analysis using BIMs	Complex structural calculations can be performed to verify, validate and test proposed designs.
Clash Detection using BIM objects in design and construction stages	An automated clash detection capability whereby the software automatically identifies where two or more components or objects interfere. Tolerances can be established to allow for build/constructability.
BIM based Collaboration and Collaborative Environments using BIMs	The sharing of single-source, unified data enables a new level of accurate collaboration by design team members. Transferring the BIM model(s) to the cloud enables real-time collaboration by remotely located design team members. The incorporated document management system enables command of version and revision control systems.
Applications using BIMs in code checking against national regulations	BIM models can be automatically verified, validated and tested against some national building code regulations.
Applications supporting the use of BIMs for Green Building Concept (*i.e.* role of: BIMs in design of Energy Efficient Buildings) Applications for Sustainable Design	The design optimisation capabilities provided by BIM can allow designs to be optimised for energy efficiency or environmental sustainability. Analysis tools enable modelling of solar gain, thermal mass and hydraulic flow. The capability to model facilitates the introduction of sustainable fabrics and technologies as their utility and efficacy can be better predicted. This can include technologies such as BiPV and ground-source heat pumps or material replacement such as rammed earth construction. Pollutants and toxins can be recorded and managed, including control of substances hazardous to health during refurbishment or eventual demolition. The proposed designs can be automatically checked against accreditation systems such as BREEAM, LEED or Malaysia's GBI (Green Building Index)
BIM and Geospatial Information Systems	This can include site technologies for setting out of works and components. RFID systems are being integrated for component location capabilities.
Creating BIM geometries through spatial information acquisition	Point cloud technology enables retrospective development of 3D BIM models.
Integration of Business Information models	ERP (enterprise resource planning) and MRP (manufacturing resource planning) integration with BIM are logical extensions. BIM models can drive manufacturing systems, for example ductwork or structural steel shop manufacturing. The entire business process can be managed and quality assured through an ERP system.
The role of BIMs in 3D City	Utilities such as power and drainage can be modelled and assessed within their

Building Information Model Application Areas	Extended Description
Models	civic setting. This enables providers and authorities to better manage urban development. Traffic flows can be modelled to improve integration and traffic planning, and transport networks can be improved at master and regional plan level. Emergency response and security/policing requirements can be visually or numerically assessed and enhanced.

Source: Developed for this book from (Eastman *et al.*, 2011, Underwood and Isikdag, 2009)

Software: Interoperability and Industry Foundation Classes (IFC)

Generally the software applications are proprietary software and operate under strict license conditions. Early attempts to introduce BIM software to the construction industry floundered owing to the inability for one vendor's software to collaborate or exchange data with another's. This phenomenon of 'interoperability' was originally addressed by industry leader Autodesk's 1994 initiative to collaborate with industry competitors in forming a *"A neutral, international and unique non for profit organisation supporting open BIM through the life cycle*" in order to "*Develop and maintain international standards for openBIM* " (buildingSMART, 2012a).

"The IFC specification is a impartial data schema. The IFC specification is developed and maintained by buildingSMART International, (formally known as International Alliance for Interoperability, IAI), and all the major software developers in this industry segment worldwide are committed to producing IFC-compliant software". The details can be described as follows (buildingSMART, 2012c):

"The buildingSMART data model:

The data schema comprises data covering the many sources from the project lifecycle. It is also registered by ISO as ISO/PAS 16739.

IFC:

It is designed to interchange BIM data between the software applications without the software having to support numerous native formats".

"IFC4:

- It was published in March 2013 with several extensions of IFC; fully integrated simple ifcXML specification; and a new documentation guide.

The exchange requirement inhibits woolliness and vagueness".

The data schema can be classified into "four conceptual layers, each individual schema is assigned to exactly one conceptual layer.

Fig. **2** shows the schema architecture.

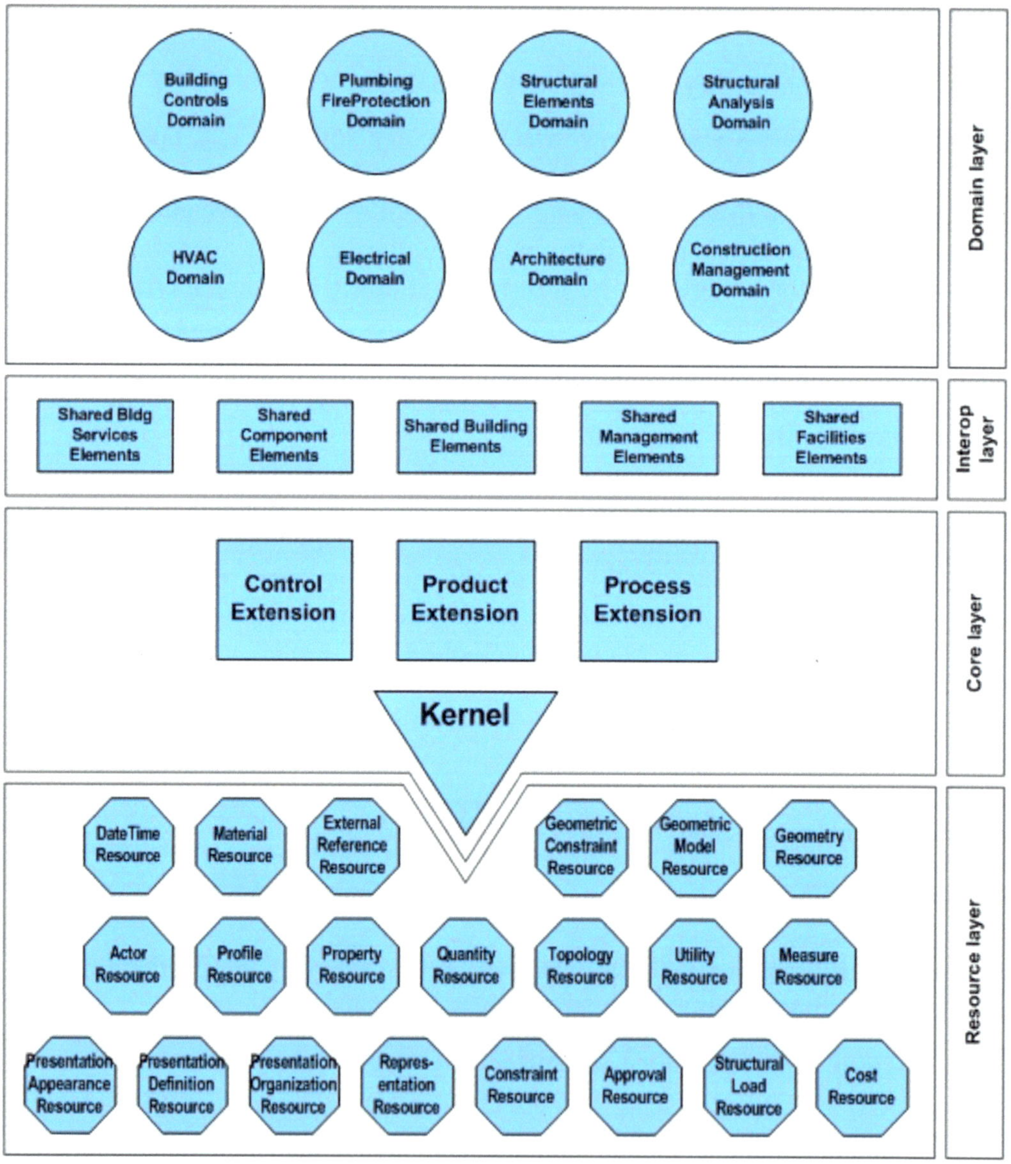

Figure 2: Data schema architecture with conceptual layers.

1. Resource layer - the bottom layer describes all individual schemas containing resource definitions;

2. Core layer - the following layer describes the kernel schema and the core extension schemas, containing the most general entity definitions,;

3. Interoperability layer - the subsequent layer describes schemas containing entity definitions that are specific to a general product, process or resource specialization used across several disciplines,;

4. Domain layer - the top layer describes schemas containing entity definitions that are specialisations of products, processes or resources specific to a certain discipline".

Software: Construction Operations Building Information Exchange (COBie)

Subsumed into many software (Autodesk, 2012a) and BIM Standards (BIM-Industry-Working-Group, 2011, National-Institute-Building-Sciences, 2012a), COBie is a data format specifically for the publication of non-geometric building model data. The specific nature, type and structure of the data is dependent upon the project phase. During design, COBie requirements focus on spaces and zones, building products, and equipment and systems. COBie requirements in the construction phase focuses on the selection, assembly and installation of components and equipment. During the commissioning phase, COBie provides verification, validation and testing benchmark data and captures the 'as-built' data.

Thus, the cumulatively aggregated data provides building asset data and information for the maintenance, management and decommissioning phases of the facilities or building. Being an open standard, Computerized Asset and Facility Management (CAFM) systems and Computerized Maintenance Management Systems (CMMS) read COBie data.

Structuring data according to COBie begins to overcome some of the system deficiencies of misaligned, inconsistent data, and unlock some of the potential of BIM for whole-life building improvement.

BIM Wash, The Hype Cycle and Technology Readiness Level

Despite the positive rhetoric from the software vendors, end-users are less convinced with interoperability between different vendors' products being considered the major barrier to successful industry adoption (Constructech, 2012), and the lack of interoperability even extending to intra-operability as consecutive versions of industry leader Autodesk's flagship BIM product, Revit, are not backward compatible (Autodesk, 2012b).

Although impressive claims are made for BIM's comprehensive power and efficacy, with some considering it a panacea for all of construction's shortcomings, others, notably Succar (2010) and Brewer *et al.,* (2011) describe a situation where the claims for BIM's potency is based too greatly on the theoretical potential as opposed to being grounded in the reality of what is currently achievable. The term 'BIM wash' has emerged to describe the over stating of BIM's pragmatic capability.

This draws parallels with Gartner's hype cycle (Gartner-Research, 2012). The hype cycle describes phases of a technology's lifecycle from innovation to its eventual common adoption. The current version of the cycle, (Gartner-Research, 2012), shown in Fig. **3** below, provides an approximation of BIM as "Big Data" and extreme information as endorsed by Constructech magazine (Constructech, 2012) and Cholakis (Cholakis, 2012), indicating that although in the early stages of development, within the next five years it should pass through a peak of inflated expectation before passing through a period of disillusionment until a period of aligned expectations and technological capability lead to a genuinely productive phase. Ahead of Big Data is Cloud/Web Platforms, which is in Gartner's 'trough of disillusionment' before being predicted to enter an upswing of productivity. The Cloud is a key enabling technology for real-time collaborative BIM modelling (BIM-Industry-Working-Group, 2011, CIB-W078, 2010, Fathi *et al.*, 2012b) and along with Big Data can expect challenging days ahead.

Sub-Section Summary

The technical platform underpinning BIM is rapidly advancing. Many successful BIM projects are recorded (American-Institute-of-Architects, 2012a). These include successful deployments beyond shared 3D modelling and include 4D chronological and 5D cost modelling instances (Kala, 2010, Hardin, 2009, McGraw-Hill-Construction, 2010, Becerik-Gerber and Rice, 2010). Solutions are also moving beyond modelling to specification preparation whereby specifications can be produced semi-automatically from models as evidenced by the UK's RIBA subsidiary, The NBS, who have created the technology (The-NBS, 2013b).

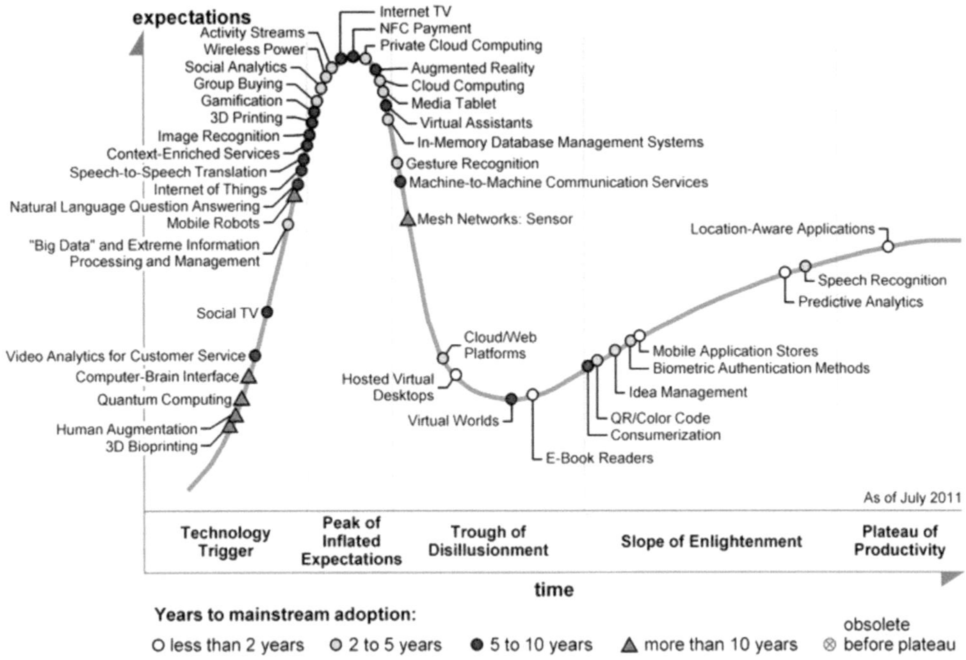

Figure 3: Gartner hype cycle 2012. (Source: Gartner-Research, 2012).

Progress has been made by both the construction industry and software vendors to integrate data and systems in open standards (buildingSMART, 2012d), as Eastman notes, "*The issues of exchange [data exchange between proprietary systems] are resolvable*" (Eastman *et al.*, 2011, p. 68). It should be noted that BIM software that can perform some of the basic parametric 3D modelling functions are available for free through free and open source software organisations, including FreeCAD (FreeCAD, 2012) and BIMserver (BIM-Server, 2012). The COBie system has been designed to be operative by basic spreadsheet software, which again is available for free. As such, there is no absolute financial constraint or software barrier to building supply chain participants to undertake a project centred on BIM.

Presently, BIM is moving to a cloud based platform (Chong *et al.*, 2014), which is likely to expand opportunities and "introduce and develop the concepts of potential innovative collaborative tools, such as Context-Aware Cloud Computing Information Systems (CACCIS), for facilitating the construction supply chain processes and networks by enhancing the opportunities for achieving better competitive advantages" (Fathi *et al.*, 2012b).

4.3.2. Process

As discussed in the preceding section, the most current BIM software is generally interoperable and integrative, however, BIM adoption must consider the wider perspective of the firm's undertakings as BIM is not solely a technical initiative.

> *"If these initiatives are to be successful it is apparent that the top tier of the TPO supply chain should share similar understanding of the technological consequences of the decision to integrate significant portions of their business processes with those of their TPO trading partners. This requires all parties to be educated, willing to be assessed for suitability, and culturally prepared for such integration"* (Brewer, 2010, p. 8).

The MacLeamy Curve, shown in Fig. **4**, is endorsed by industry authorities, such as buildingSMART Alliance (MacLeamy, 2007), and illustrates the fundamental nature of the process change, whereby the entire design process is brought to the front end of the project in keeping with BIM's identity as a prototyping tool. The early work effort ensures that the design is completed and verified prior to the construction phase (or the sequential construction phase on fast-track projects).

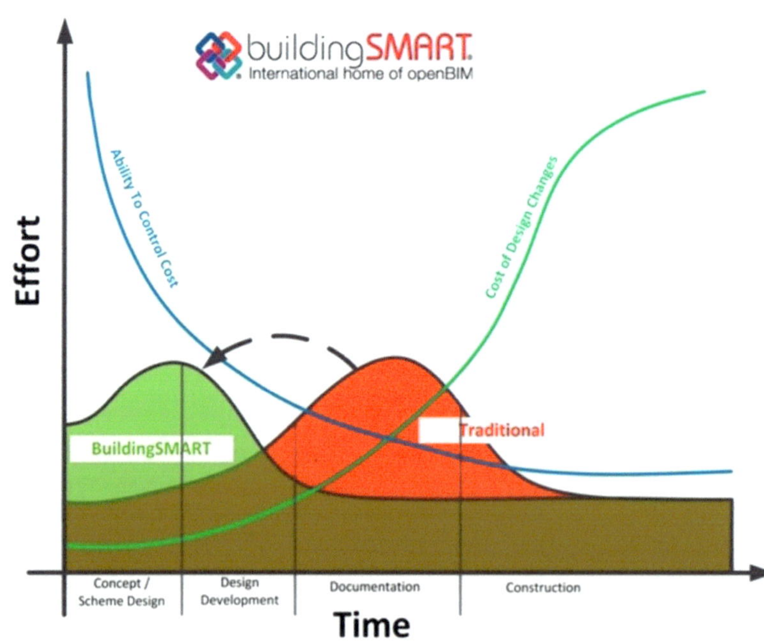

Figure 4: The MacLeamy curve. (Source: MacLeamy, 2007).

In addition to chronological changes in design activity, there are inherent changes to the design activity itself. The change from 2D design to 3D design requires change in technique and approach (Shahrin *et al.*, 2010, Yin *et al.*, 2009). Often this requires the conversion, through modeller's effort, of 2D drawings to 3D model (Senescu and Haymaker, 2009) as most designers do not work in 3D. In turn, the 3D models typically need to be reduced back down to 2D drawings for site, communication, record and contract purposes (Khanzode, 2010).

In addition to the internal impacts upon a firm's design processes, Brewer *et al.*'s (2011, p. 247) Australian case study highlights the wider, external impacts:

> *"This case study revealed a number of the paradoxes inherent in the industry at this time. Firstly, a number of the project participants had invested in BIM-capable technologies, and although each was satisfied with their use of it at an intra-firm level, the same could not be said for inter-firm information sharing. However none of these firms regarded this as an ICT failure, perhaps because none of them had genuinely BIM-ready business processes, which are inherently boundary-spanning, and require a degree of shared business processes".*

Brewer *et al.* (2011, p. 247) identify the type and nature of inter-firm collaboration issues:

> *"... the beneficial effects of ICT-mediated collaboration [BIM] on this project were realised in spite of the procurement mechanisms, not because of them. They arose because the specialist subcontractor was prepared to set aside the letter of the contract and pursue a more collaborative approach in order to achieve the project objective.This required them to be exposed to what amounted to unregulated risk and unrewarded risk-taking, a behaviour that would be considered unusual in the construction industry... This highlights both the inadequacy of existing procurement mechanisms for suitability in the BIM age, and the necessity of developing new mechanisms to fill this gap, if BIM is to fulfil its potential".*

To use BIM effectively the communication between different stakeholders in the construction process needs substantial development. When required information is available at the time of need and the quality of information is adequate, the construction process will be considerably enhanced.

To address the issues of process failings, buildingSMART, the international authority on BIM has developed an appropriate process model. The buildingSMART standard provides a common understanding for all the parties when to exchange information and precisely what is needed. This has been achieved in collaboration with firstly, the British Standards Institute (BSI) and then further ratified and adopted by the International Standards Office (ISO). Together they have produced a range of standards, with notable ones described below.

BuildingSMART

BuildingSMART International is the leading, global body guiding the development of open, internationally accepted BIM standards, tools and training to encourage the widespread uptake of open BIM by the Architecture, Engineering & Construction (AEC) and Facilities Management (FM) industries.

BuildingSMART International is a neutral, not-for-profit organisation driving the global use of BIM (Building Information Modelling). It is structured in Regional Chapters - each representing a country or group of countries performing together. Each Regional Chapter is a distinct organisation and is formed according to indigenous custom. Malaysia is in the process of forming its own chapter, and MY BuildingSMART Berhad is intended to be the Malaysian Chapter of BuildingSMART International.

Fig. **5** describes the high level activity areas considered by buildingSMART.

Central to the process dimension of successful BIM implementation is the progressive flow of information throughout the project and building's lifecycle. At critical junctures, information is exchanged by participants and this 'Information Exchange' is ongoing throughout any building project. At times the exchange is formal but frequently it is not. Formal exchange conforms to contractual obligations, while informal exchange that supports on-going design development may take place on a daily basis as part of normal working practices.

An important objective of BIM is the ultimate development of a Project Information Model. This will include all the information (documents, graphical models and non-graphical data) needed for the design and construction of an asset. PAS 1192-2 provides a structure around its delivery and captures the information *via* information delivery cycle (IDC). The IDC encompasses both informal and formal exchange points. The formal exchanges are termed

'Employer's Decision Points' - the important junctures in the delivery processes where the employer requires specific information to support activities pertaining to the project. The timing and nature of these will be particular to each project, however each project will progress through a planning, design and construction process prior to handover.

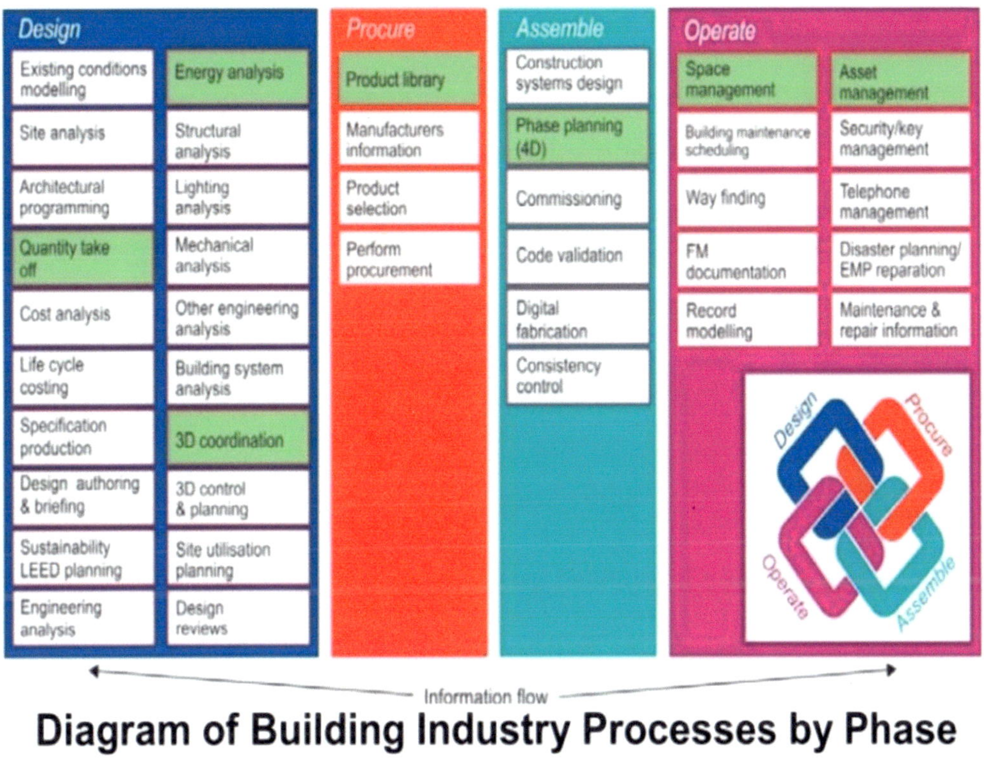

Figure 5: High level activity areas. (Source: buildingSMART 2012a).

PAS 1192-2 is a Publically Available Specification (PAS) that provides a framework for achieving building information modelling (BIM) Level 2. The requirements extend the existing code of practice for the collaborative production of architectural, engineering and construction information, defined in BS 1192:2007. BS 1192 provides a methodology for managing the production, distribution and quality of construction information, using a structured process for collaboration and a specified naming policy which aims to help different designers to reach compatible stages of design work at the same time (British-Standards-Institution, 2008a).

The IDC is shown in Fig. **6**.

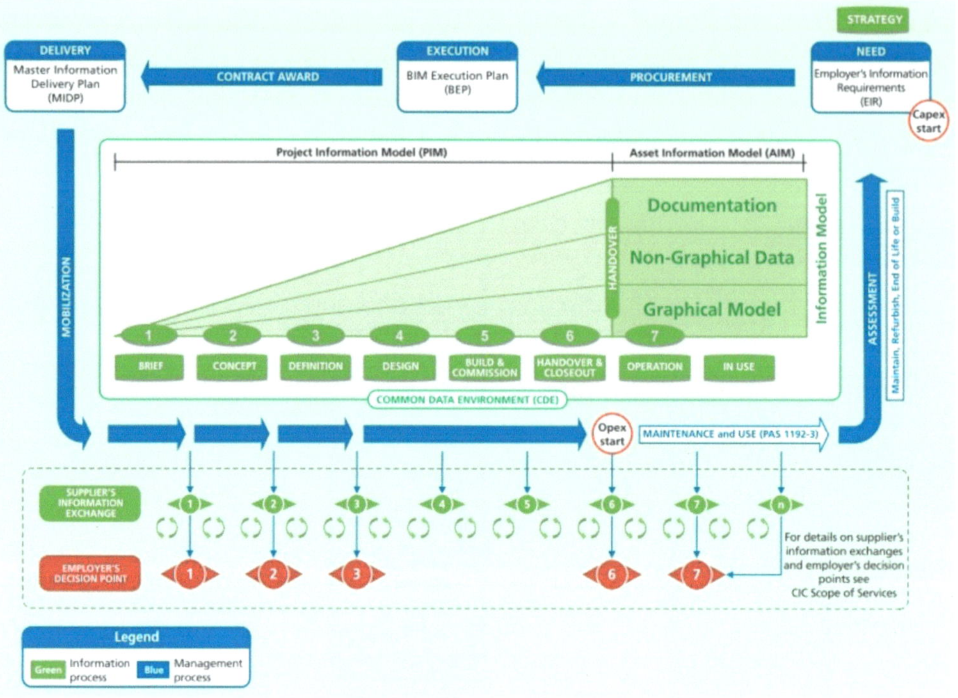

Figure 6: Information delivery cycle. (Source: PAS 1192-2).

The following explanation is provided by PAS 1192-2:

> *"NOTE 1 - The information delivery cycle as shown in Fig.* **6** *has two distinct points of entry. For stand-alone new-build projects, start at the top right box "Need", but for projects that are part of a larger portfolio or estate, or for projects working on existing buildings and structures, then start at the right-hand arrow "Assessment" which draws on the information in the existing AIM".*

> *"NOTE 2 - The information delivery cycle shows in BLUE the generic process of identifying a project need (which may be for design services, for construction or for supply of goods), procuring and awarding a contract, mobilizing a supplier and generating production information and asset information relevant to the need. This cycle is followed for every aspect of a project, including the refinement of design information through the seven project stages shown in GREEN".*

> *"NOTE 3 - The GREEN numbered ovals and annotated lozenges refer to the CIC Scope of Services stages. The GREEN image represents the CDE*

that will collect, manage, disseminate, exchange and retrieve information through the lifecycle".

"NOTE 4 - Information exchanges between project team members are indicated by small GREEN balloons".

"NOTE 5 - Information exchanges between the project team and the employer are indicated by larger red balloons to answer the Plain Language questions posed by the employer defined in the employer's information requirements".

"NOTE 6 - Copyright is claimed in this illustration by Mervyn Richards. Reproduction of this illustration and making products from it might infringe that copyright".

Consistent with the IDC shown above, BuildingSMART has produced a standardised process, previously named Information Delivery Manual (IDM). This has been further developed in collaboration with the International Standards Office (ISO) as ISO 29481-1:2010. It specifies when particular sorts of information are needed throughout the building lifecycle. This stipulates the information that a specific participant (architect, engineer or contractor) requires at key junctures of the process. It may include information that is necessary for activities including; cost estimating, volume of materials and time planning.

"ISO 29481-1:2010 specifies a methodology that unites the flow of construction processes with the specification of the information required by this flow, a form in which the information should be specified, and an appropriate way to map and describe the information processes within a construction life cycle.

It defines how to recognise and describe the processes performed within construction, and the information essential for their execution and the satisfactory achievement. It also describes how the information can be additionally detailed to facilitate solutions provided by BIM system providers in a manner that empowers its reuse.

ISO states its objectives as:

"The IDM will provide the integrated comprehensive reference required by identifying the discrete processes undertaken within building construction together with the information that is required for and results from their execution.

The IDM will target both BIM users and solution providers.

- *For BIM users, it will provide a simple to understand, plain language description of building construction processes, the requirements for information to be provided to enable the process to be carried out successfully, additional information that may need to be provided by the user and the expected end results of the process.*

- *For BIM solution providers, it will identify and describe the detailed functional breakdown of the process and the IFC capabilities needing to be supported for each functional part in terms of the entities, attributes, property sets and properties required.*

Additionally, the IDM will provide an improved methodology for development of future extensions to the IFC model that will include close descriptions of business requirements and IFC support so as to support accelerated implementation".

The adoption of BIM is shown as fundamentally altering the operational processes of a firm and its allied supplied chain, including the project and design teams.

BIM Standards and BIM Execution Plans

There are a range of BIM Standards (AEC-(UK), 2012c, Indiana-University, 2012, National-Institute-Building-Sciences, 2012a, New-York-City-Department-of-Design-&-Construction, 2012, Singapore-Gov, 2012b, The-NBS, 2012b) and support organisations that provide guidance, such as BIM-Task-Group (2013). These provide detail of how a firm can organise its resources to achieve BIM capability. They variously describe best-practice strategic, tactical and operational processes and procedures for managing BIM production. In so doing, they provide a common basis for collaborative working with other BIM project participants. They also describe methods for collaborative working between project participants, however they are very much a work-in-progress (Berard and Karlshoej, 2012). The American Institute of Architects has produced a survey of thirty seven BIM Standards and Guidelines (American-Institute-of-Architects, 2012e) that includes some cursory information on each, however there is considerable scope for analysis, leading to convergence, which may provide a suitable point of departure for further research.

Level of Detail/Development

BIM Standards and Execution Plans provide guidance on the process whereby collaborative teams can produce BIM models (Shou *et al.*, 2014). The specification richness and utility increase as the project progresses and successive information is added to the model. Each construction project is unique and consequently has unique requirements for its BIM model(s). As advised by the BIM Standard, the purpose of the model should be agreed and this will help establish the requisite level of data or information to be specified for the BIM model. This has been termed 'Level of Detail' (AEC-(UK), 2012c) or 'Level of Development' (LOD). As the project progresses, the model will similarly progress through successive LODs, with each LOD describing the uses for which the state of the model is appropriate. Fig. **7** provides a matrix of development level and uses. The LOD concept has found wide acceptance in the global industry, being used in BIM Standards and contracts to aid communication.

Level of Detail ->	100	200	300	400	500
Model Content					
Design & Coordination (function / form / behavior)	Non-geometric data or line work, areas, volumes zones, etc.	Generic elements shown in three dimensions - maximum size - purpose	Specific elements Confirmed 3D Object Geometry - dimensions - capacities - connections	Shop drawing/ fabrication - purchase - manufacture - install - specified	As-built - actual
Authorized Uses					
4D Scheduling	total project construction duration phasing of major elements	Time-scaled, ordered appearance of major activities	Time-scaled, ordered appearance of detailed assemblies	Fabrication and assembly detail including construction means and methods (cranes, man-lifts, shoring, etc.)	
Cost Estimating	Conceptual cost allowance Example $/sf of floor area, $/hospital bed, $/parking stall, etc. assumptions on future content	Estimated cost based on measurement of generic element, e.g., generic interior wall.	Estimated cost based on measurement of specific assembly, e.g., specific wall type.	Committed purchase price of specific assembly at Buyout.	Record costs
Program Compliance	Gross departmental areas	Specific room requirements	FF&E, casework, utility connections		
Sustainable Materials	LEED strategies	Approximate quantities of materials by LEED categories	Precise quantities of materials with percentages of recycled/locally purchased materials	Specific manufacturer selections	Purchase documentation
Environmental: Lighting, Energy Use, Air Movement Analysis/Simulation	Strategy and performance criteria based on volumes and areas	Conceptual design based on geometry and assumed system types	Approximate simulation based on specific building assemblies and engineered systems	Precise simulation based on specific manufacturer and detailed system components	Commissioning and recording of measured performance

Figure 7: Level of detail/development. (Source: AEC-(UK), 2012).

Additional Process Support

A number of initiatives have been attempted to support the alignment of deliverables and coordination of production information between design and management companies collaborating on construction projects, examples of which are described in brief below:

Avanti

The core of the Avanti method is an approach where all geo-spatial information is generated with the same origin point or master datum (0,0,0). It is supported by principles of providing early access to all project information by all stakeholders, early involvement of the supply chain, and sharing of all information in an agreed and systematic manner (Constructing-Excellence, 2012).

Sub-Section Summary

The literature identifies that a large array of tools, guides, protocols, procedures, standards and execution plans exist to support adopting organisations and project teams in their use of BIM. The documents generally state that they are works-in-progress and are expected to be updated in line with developments in technology and process knowledge.

The next sub-section addresses the 'people' component of BIM.

4.3.3. People

An infusion of new technology and new operational processes requires a complementary change in human resource capability (Eastman *et al.*, 2011, Gu and London, 2010, Hardin, 2009, Samuelson and Björk, 2011, Sebastian, 2011, Singh *et al.*, 2011, UK-Gov, 2011a) with authorities concurring that this necessitates the new, technical role of 'model manager' whose function is to manage access, availability, coordination, configuration, consistency and change management of the model for team participants. In this early phase of BIM use, it is envisioned that the role, additional to clear technical competency, includes a mentoring and leadership quality to induct legacy team members into a new working paradigm (Dossick and Neff, 2010, Arayici *et al.*, 2011).

Beyond the technical specialism of model manager, team members with process and procedural expertise, referred to as 'modellers with construction experience' by Eastman (2011, p. 355), may be required depending upon the scale, complexity, level of collaboration, experience and capability of the design team members. At the minimum, training and technical support will be required for any significant changes in work process, procedure or tools technology.

4.4. MATURITY MODELS

The above sections describe the technical, process and human resource perspectives on BIM. An increase in any one perspective will produce a marginal increase in BIM capability but there is a symbiotic dependency on the other two perspectives to attain significant improvement. This relationship is expressed by one perspective of the Bew-Richards maturity model Fig. **8**. Endorsement of these levels is provided by the UK government (UK-Gov, 2011a) who require that all government projects achieve BIM level 2 capability by 2016 (UK-Gov, 2011b).

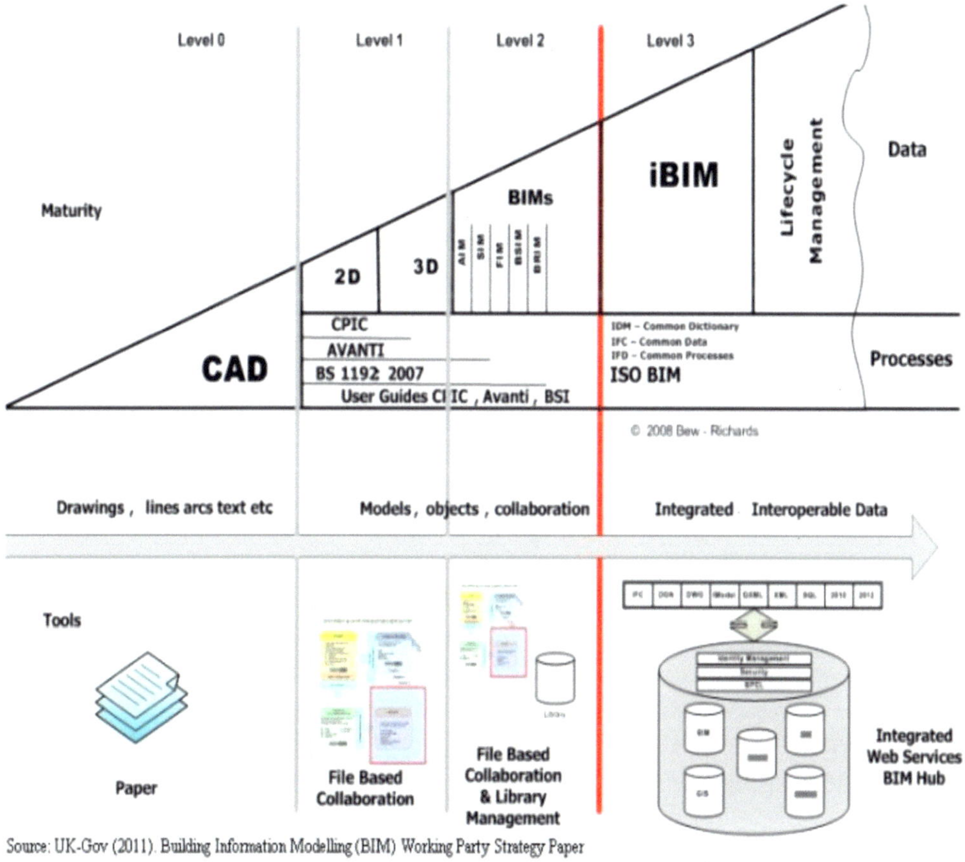

Source: UK-Gov (2011). Building Information Modelling (BIM) Working Party Strategy Paper

Figure 8: Bew-Richards BIM maturity model.

The maturity model describes growing sophistication whereby as the BIM capability matures, it is possible to differentiate between, and target design for, optimisation of CAPEX (capital expenditure) or OPEX (operating expenditure) of the facilities. This would be consistent with the theoretical 1:5:200 relative cost ratio. The inclusion of *Construction Project Information Committee (*CPIC), Avanti and BS drives convergence of a practice's processes towards consistent, standardised processes that will produce homogenous models that can be shared. Industry-wide adoption of these standards, or similar standards, facilitates the spread of BIM across the industry.

The Collaborative BIM Decision Framework

Gu and London (2010, p. 994) utilise the Royal Australian Institute of Architects maturity model for the development of their Australian context Collaborative BIM Decision Framework. The framework provides a consecutive maturity path that progresses with the addition of actors, tools and technical capability as follows (Gu and London, 2010):

> *"Level 0: CAD (Computer-aided Design) based (2D and 3D) - design disciplines who are designing, documenting and creating visualisations but who have not yet fully embraced object-oriented modelling and the concept of embedded information and/or appended/linked object information".*
>
> *"Level 1: modelling - single disciplinary use of object-oriented 3D modelling software within one discipline".*
>
> *"Level 2: collaboration - sharing of object-oriented models between two or more disciplines".*
>
> *"Level 3: integration - integration of several multidisciplinary models using BIM model servers with the ultimate aim of moving from local servers to a web based environment".*

Interactive BIM Capability Maturity Model

The American National Institute of Building Sciences' National Building Information Model Standard (NBIMS) suite of documents and tools contains an Interactive BIM Capability Maturity Model (National-institute-Building-Sciences-

and-buildingSMART-Alliance, 2012). The model assesses BIM maturity across eleven dimensions as provided in Table **10** below:

Table 10: NIBS NBIM interactive capability maturity model

Dimension	Description
Data Richness	"Identifies the completeness of the building Information Model from initially very few pieces of unrelated data to the point of it becoming valuable information and ultimately corporate knowledge about a facility".
Life-cycle Views	"Views refer to the phases of the project and identifying how many phases are to be covered by the BIM. One would start as individual stove pipes of information and then begin linking those together and taking advantage of information gathered by the authoritative source of the information. This category has high cost reduction, high value implications based on the elimination of duplicative data gathering. The goal would be to support functions outside the traditional facility management roles, such as first responders".
Roles Or Disciplines	"Roles refer to the players involved in the business process and how the information flows. This is also critical to reducing the cost of data re-collection. Disciplines are often involved in more than one view as either a provider or consumer of information. Our goal is to involve both internal and external roles as both providers and consumers of the same information so that data does not have to be re-created and that the authoritative source is the true provider of the information".
Business process	"Change Management identifies a methodology used to change business processes that have been developed by an organization. If a business process is found to be flawed on in need of improvement, one institutes a "root cause analysis" of the problem and then adjusts the business process based on that analysis. Since this is related to the following item, business processes it should come after it".
Delivery Method	"The business process defines how business is accomplished. If the data and information is gathered as part of the business process then data gathering is a no cost requirement. If data is gathered as a separate process then the data will likely not be accurate. The goal is to have data both collected and maintained in a real time environment, so as physical changes are made they are reflected for others to access in their portion of the business process".
Timeliness/ Response	"While some information is more static than other information it all changes and up to the minute accuracy may be critical in emergency situations. The closer to accurate real time information you can be the better quality the decisions that are made. Some of those decisions may be life saving in nature".
Change Management	"Data delivery is also critical to success. If data is only available on one machine then sharing can not occur other than by email or hard copy. In a structured networked environment if information is centrally stored or accessible then some sharing will occur. If the model is a systems oriented architecture (SOA) in a web enabled environment the nentcentricity will occur and information will be available in a controlled environment to the appropriate players. Information assurance must be engineered into all phases".
Graphical Information	"Often the starting point is a non-graphical environment. The advent of graphics helps paint a clearer picture for all involved. As standards are applied then information can begin to flow as the provider and receiver must have the same standards in place. As 3D images come into play more consumers of the information will have a common view and a higher level of understanding will occur. As time and cost are added then the interfaces

Dimension	Description
	can be expanded significantly".
Spatial Capability	"Understanding where something is in space is significant to many information interfaces and the richness of the information. Energy calculations must know where the heat gains will come from, first responders need to know where water supplies and utility cutoffs are located in relation to the facility".
Information Accuracy	"Having a way to ensure that information remains accurate Is only possible through some mathematical ground truth capability. Having a mathematical product will also allow for better management by supporting difficult to game metrics. These numbers can be used for occupancy, information collection completeness and overall inventory calculations".
Interoperability/ IFC Support	"Our ultimate goal is to ensure interoperability of information. Getting accurate information to the party requiring the information. There are many ways to achieve this, however the most effective is to use a standards based approach to ensure that information is a form that it can be shared and products are available that can read that standard for of information".

Source: (National-institute-Building-Sciences-and-buildingSMART-Alliance, 2012)

Significantly, this model is provided by the American Government's National Institute of Building Sciences, and specifically the National Building Information Model Standard (NBIMS) Committee, as a collaborative and jointly published effort with buildingSMART Alliance, who itself represents the dominant software developers, such as Autodesk, Bentley, Nemetschek Vectorworks, Graphisoft and Trimble Navigation (recent purchasers of SketchUp from Google).

4.4.1. Discussion on Maturity Levels and What Constitutes BIM

The dimensions of the NIBS model provide a broader and more comprehensive spectrum against which BIM capability can be measured than the UK or Australian model. The model encourages users to rate their capability (or desire) on a 1-10 scale, which although not perfectly linear, enables smoother, or more granular, progression than the 3 step-change models of the UK or Australia. On the other hand, the comprehensiveness and granularity necessitate the inclusion of complex concepts that make the model cumbersome and difficult to comprehend.

With regards to what the providers of these models advise as constituting BIM, academic differences appear. The UK bodies consider that their Level 2 indicates true BIM capability. The approach may utilise 4D programme data and 5D cost elements as well as feed operational systems.

The Australian bodies consider that their Level 2 signifies true BIM capability:

Level 2: collaboration - sharing of object-oriented models between two or more disciplines.

The American bodies, in keeping with their Interactive Capability Maturity Model, provide a description according to their model (National-Institute-Building-Sciences, 2012b), namely, "Spatial Capability, Roles or Disciplines, Data Richness, Delivery Method, Change Management or ITIL Maturity Assessment, Business Process, Information Accuracy, and Lifecycle Views"

The UK's is notable in that it seeks integration between BIM models and an ERP system. In comparison, the Australian model is more implied by requiring the sharing of models across disciplines. Inherently this requires aligned model production processes and standards. The American model also emphasises the importance of cross-discipline BIM model sharing. With regards to internal integration, as of its latest version in May 2012, its sections on 'Change Management' and 'Business Process' are blank, this is NIBS and buildingSMART's omission, not researcher's. Thus it is inconclusive. However, it should be noted that considerable collaboration and alignment is taking place between the UK and American bodies, (National-Institute-of-Building-Sciences, 2012).

Although there is some disagreement between the authorities on what constitutes BIM, the fundamental similarity is the ability to share models between disciplines, typically Architectural, Structural and MEP (mechanical, electrical, plumbing). Thus, the common component of making data available for sharing with teams, internal or external, similar or disparate, remains, and this should be observable both in action to indicate BIM's existence, or in awareness, to indicate complimentary understanding of what BIM is.

In alignment with this is the production of legal and contractual advice and documentation from industry organisations. Notable are the American-Institute-of-Architects (2012d) and the UK's Construction-Industry-Council (2013) that provide legal frameworks for BIM projects. Notably, they are both intended to support projects to BIM Level 2 and maintain firm boundaries between design disciplines. These documents emanate from industry concerns over legal responsibilities and rights that may be impacted by working in a collaborative BIM environment.

Section Summary

BIM, as an aggregation of Building Information Models, Building Information Modelling and Building Information Management (buildingSMART, 2012b), impacts adopting firms across the dimensions of people, process and technology (Grilo and Jardim-Goncalves, 2010). It requires and produces a fundamental change in which the organisation functions (Sacks and Barak, 2008). The convergence of the dual triumvirates of; 'people, process and technology' and 'model, modelling and management' are illustrated by Gu and London's (2010) scoping activities, purpose and phase matrix as presented in Fig. **9**. Firstly, it describes some strategic steps towards the construction of project BIM models that are shared between disciplines. Importantly, it establishes the linkage between BIM models and project phases, thus ensuring that valuable information is available at a valuable time. It identifies senior management buy-in as a prerequisite and identifies the enabling function of collaborative document management systems. Secondly, the matrix illustrates some of the potential range of purposes that BIM model information can be used for. Again, this is linked to the project phases to ensure that information availability is timely.

BIM inherently requires a collaborative environment in which to operate (Eastman *et al.*, 2011). The authorities concur that 'minimum BIM' requires sharing of models between different disciplines (AEC-(UK), 2012a, American-Institute-of-Architects, 2012b, Arayici *et al.*, 2012, Autodesk, 2011, buildingSMART, 2012b, buildingSMART-Australasia, 2012). By definition, BIM cannot be adopted in isolation by the AECO sector. BIM can only be adopted in concert with other key stakeholders (George, 2012). Following Saxon (2005) and Gu and London (2010) the AEC industry may include other design team members, client bodies, developers and contractors and all forms of Government.

BIM has the potential capability to be considered a disruptive technology (Bower and Christensen, 1995) whereby it can supplant existing technology, such as 2D CAD. Certainly issues regarding BIM's application in real-world environments suggest the immature technology lacks refinement and has performance problems that are the hallmark of disruptive technology. As pan-project BIM capability matures, the benefits of 4D, 5D, 6D modelling will emerge (Sacks *et al.*, 2010a) - although these could be considered as the lesser, sustaining technologies that have evolved from existing technologies - however, BIM is a collaborative platform that facilitates convergence of technologies, and it is in these nD dimensions that BIM has the capability to break the status quo and revolutionise the AECO sectors, and by extension, through creative destruction, bring about a *"process of industrial mutation that incessantly revolutionizes the economic structure from within, incessantly destroying the old one, incessantly creating a new one"* (Schumpeter, 1994 [1942], p. 140).

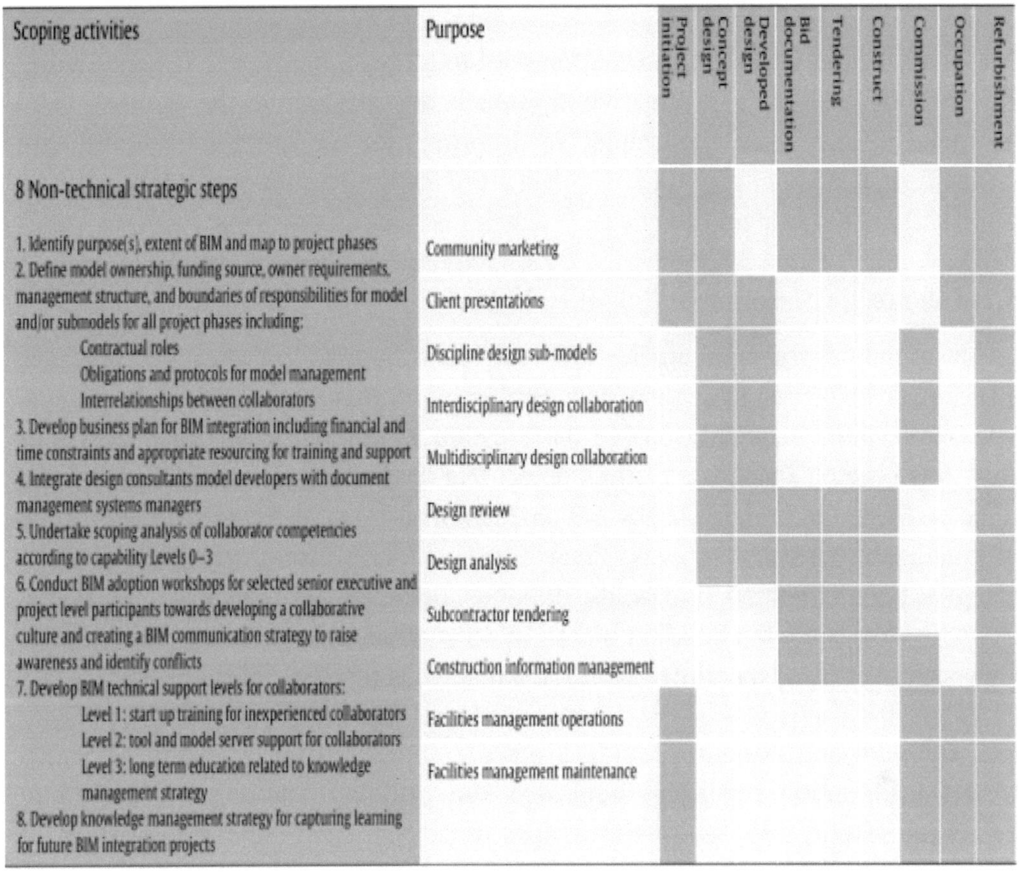

Figure 9: Scoping activities, purpose and phase matrix. (Source: Gu and London (2010)).

4.5. BIM SOFTWARE

4.5.1. Introduction and the Concept

This chapter begins with a brief historical perspective on BIM software. It then proceeds to describe current and commonly available BIM software. It concludes with an overview of training programmes that are pertinent to the obtaining of BIM software skills.

Developments in computer and communication systems in Architecture, Engineering and Construction have evolved to enrich 3D CAD models of buildings and structures with complementary data such as physical characteristics, unit costs, quantity take-offs and functional characteristics. This methodology became known as the Building Information Modelling.

BIM serves as an integrated information pool to all team members. The modelling tools provide designers a creative approach for designing the project. The construction team uses the model to improve the constructability and site coordination. BIM is intended to be a data store for the facility operator to use and maintain along the project lifecyle.

4.5.2. Geometric Shapes and Parametric Modelling

A wide range of software applications can contribute to the creation, analysis, enhancement and communication of a Building Information Model for a project. Some applications may only enable the creation of geometric or basic shapes or solids. A geometric shape is a distinct object that has an external surface or specific outline or form. Geometric shapes are characterized by straight, precise edges and mathematically consistent curves. Such objects include circle, triangles, squares, rectangles, parallelograms, trapezoids, rhombus, octagons, pentagons and hexagons. Geometric shapes are the building blocks of most structures, including buildings, in the physical world and the virtual world in which BIM models reside.

Some software applications provide a more comprehensive authoring of multi-attribute information regarding buildings and building components, including information about how geometric shapes relate to each other. According to Eastman *et al.*, (2011), a building model configuration is *"defined by the user as a dimensionally-controlled parametric structure, using grids, floor levels, and other project references planes. Alternatively, these can simply be floor planes, wall centre-lines or a combination of them. With these embedded object instances and parametric settings, the model configuration defines an instance of the building"*. This enables rules, or parameters, to be encoded within the software, so that a change in one geometric shape will execute a change on a related shape.

Parametric object modelling provides a powerful way to generate and modify geometry. Parametric modelling is the critical productivity capability that allows design changes to update automatically throughout the entire model. Without it, model based design generation would be awkward and error-prone. It is likely that

3D modelling would not be commercially viable for building design and construction without the automatic update feature made possible by parametric capabilities. Each BIM software application differs with regard to the parametric object families it provides, the rule embedded within it, and the resulting design behaviour.

Currently, no single software provides a complete, end-to-end capability for all use or user needs. Instead, a variety of software applications are used through the lifecycle and these are determined by the project and user needs and an example is presented in Table **11** below.

Table 11: BIM plan of work

	Appraisal and Brief Development	**Design**	**Pre-Construction**	**Construction**	**In-Use**
	RIBA Stage A-B	*RIBA Stage C-E*	*RIBA Stage F-H*	*RIBA Stage J-K*	*RIBA Stage L*
BIM Deliverable	"Macro BIM and Optioneering: • Production of rapid models to facilitate analysis of multiple design concepts, in order to establish feasibility and develop the brief. • These sketch models linked to environmental and cost data reduce risk and confirm viability".	"Develop the BIM to include: • Conceptual design of the architecture, structural and building services systems. • Outline specification and embedded cost data, in sufficient detail to fix the project brief and obtain planning permission. • Including the integration of information for statutory standards and construction safety".	"Refine the BIM into a virtual prototype: • To include co-ordinate design of building elements in sufficient detail to enable procurement and subsequent construction of trade packages. • Discharge outstanding statutory and code compliance issues. • BIM to include cost (5D) and programme (4D) information, to allow real time value engineering and programme simulations".	"Revise the BIM to reflect site amendments and as built information in order to: • Become a source of construction and co-ordination information to site team. • Allow alternative programmes to be run to simulate and analyze the potential impact of changes • Allow simulations to be run to reduce waste on-site and ensure products are delivered when needed and not stored on site".	"Complete the BIM in order to provide: • An accurate source of information about the as-built spaces and systems to facilitate the management and operation of the building. • Creating a single coordinated and internally consistent database for the building".

	Appraisal and Brief Development	Design	Pre-Construction	Construction	In-Use
	RIBA Stage A-B	*RIBA Stage C-E*	*RIBA Stage F-H*	*RIBA Stage J-K*	*RIBA Stage L*
Tools	Autodesk Revit Architecture Autodesk Ecotech Autodesk Revit MEP Autodesk Project Vasari Beck Technologies D Profiler Onuma System Vico Office Suite	Autodesk Revit Architecture Autodesk Revit MEP Autodesk Revit Structure Autodesk Ecotech Autodesk Navisworks Manage Autodesk Quantity Takeoff Autodesk 3ds Max Design Vico Office Suite	Autodesk Revit Architecture Autodesk Revit MEP Autodesk Revit Structure Autodesk Ecotech Autodesk Navisworks Manage Autodesk Quantity Takeoff Autodesk Inventor McNeel Rhino3D Vico Office Suite	Autodesk Revit Architecture Autodesk Revit MEP Autodesk Revit Structure Autodesk Ecotech Autodesk Navisworks Manage Autodesk Quantity Takeoff Vico Office Suite	Autodesk Navisworks

Source: Developed for this book from (David Miller Limited, 2011).

The illustration describes the lifecycle of the project, using the RIBA Plan of Work (Royal-Institute-of-British-Architects, 2013 #969), and supplements it with a BIM overlay. The overlay identifies the deliverables from the BIM process that will add value to each stage of the project lifecycle. In so doing, it identifies the requisite inputs and outputs and the necessary tools, as software applications, that will enable the process. Importantly, it also describes the workflow process and which software application is used to achieve the output.

4.5.3. Building Information Modeling Software

At present, a range of software applications are available to perform 3D modelling. However, not all have genuine BIM capability. From a technical perspective, 3D modelling software can be categorised into two typologies, surface modelling and solid modelling tools. The surface modellers are software with 3D capability but without parametric value in the generated models, while solid modellers are 3D modellers that can embed the model with rich parametric data and capabilities that will enable the model to depict the real world project.

4.5.4. Software Use by Vendors in Selected Countries

No precise data exists with regards to BIM software being used, however, both the UK (The-NBS, 2012a) and New Zealand (Masterspec, 2012) undertook recent national BIM surveys that are likely indicative of global usage. The provided statistics are misleading as they mix software tools, such as Sketchup, that are not capable of parametric modelling, and by extension unable to produce BIM models, with tools that are capable of parametric modelling and hence capable of producing BIM models. Furthermore, it is common for design practices to use the software of more than one vendor, so that an Architect may well use Sketchup and Revit simultaneously.

Table **12** below suggests that Autodesk are the dominant AEC industry software provider.

Table 12: CAD and BIM software use in selected countries

Software	UK	New Zealand	Malaysia (estimated)
Autodesk AutoCAD	28%	19%	60%
Autodesk AutoCAD LT	22%		
Autodesk Revit	17%	21%	12%
Nemetschek Vectorworks	9%	10%	3%
Graphisoft ArchiCAD	8%	37%	4%
Other	7%	8%	8%
Sketchup (Trimble or Google)	4%	3%	10%
Bentley Microstation	3%	1%	1%
Bentley Building Suite	1%		1%
Nemetscheck Allplan	1%		1%

Source: Developed for this book

Selected BIM authoring software are described in Table **13** below:

Table 13: Selected BIM authoring software

Product Name	Vendor	BIM Use	Vendor's Description	Primary Function
Revit Building Design Suites - includes Revit Architecture, Structure and MEP	Autodesk	Architecture, Structure, MEP and site design	"Revit is specifically built for BIM, empowering design and construction professionals to bring ideas from concept to construction with a coordinated and consistent model-based approach. Revit is a single application that includes features for architectural design, MEP and structural	Architectural, Structural, MEP Modeling and parametric design.

Product Name	Vendor	BIM Use	Vendor's Description	Primary Function
			engineering, and construction".	
Bentley BIM Suite - Includes MicroStation, Bentley Architecture, Bentley Structural, Bentley Building Mechanical Systems, Bentley Building Electrical Systems, Bentley Building Electrical Systems for AutoCAD, Generative Design and Generative Components	Bentley Systems	Multi-discipline	"Each discipline-specific application provides an informed work environment to support the design and documentation process throughout all phases of the project lifecycle-from conceptual design and construction documentation to coordination and construction and allow distributed teams to"build as one" within a managed environment. Building Information Modeling (BIM) is a way of approaching the design and documentation of building projects - by modeling and managing not just graphics, but also information. This information allows the automatic generation of drawings and reports, design analysis, schedule simulation, facilities management, and more - ultimately enabling the building team to make better-informed decisions and to produce better buildings. Generative design enables architects and engineers to pursue designs and achieve results that were virtually unthinkable before. Using associative parametrics and computational methods, designers can explore a broad range of "what-if" alternatives for even the most complex buildings, quickly and easily. Prerequisites: MicroStation® v8.5.2 or higher, Microsoft Word & Excel (for reporting), Supports DGN,DWG,DXF, PDF, IFC, SKP,3DS, CIS2,gbXML and more".	"Architectural, Structural, Mechanical, Electrical and Generative Components - all within the 3D modeling environment"
SketchUp Pro	Trimble and Google	Multi-discipline	SketchUp Pro is used to quickly create accurate 3D models for pursuit and	3D Architectural and Structural modelling

Product Name	Vendor	BIM Use	Vendor's Description	Primary Function
			marketing, preliminary estimation, detailing, site logistics and staging, design and construction validation, sequencing and line-of-sight analysis.	
ArchiCAD 3D	Graphisoft	Architecture, MEP and site design	ArchiCAD was the first object oriented BIM Architectural application available in the commercial market, and the only BIM architectural application running on the MAC platform, as well as Windows	Architectural Modeling
Vectorworks	Nemetschek	Architecture	From design concept to construction documentation and every design phase in between, Vectorworks Designer marries precision 2D drafting and flexible 3D modeling with state of the art technology.	3D Architectural Modeling

Source: Adapted for this research from {BIM-Forum, 2011 #974}

Owing to its current dominance in the industry, the following section provides a deeper review of Autodesk's Revit.

4.5.5. Autodesk Revit

Specifically developed for building information modelling (BIM), Revit helps designers develop design concepts and forms, and better maintain design data through documentation and construction. This enables essential BIM data to be shared with project partners for a more collaborative, integrated building design process. In turn, this supports a more efficient design process that enables enhanced design analysis, clash detection, construction planning, and material production.

The Revit interface enables users to control entire buildings or components of a project in the project environment, and discrete geometric shapes in the family editor environment. Revit's authoring tools can manipulate pre-made solid objects or imported geometric models.

Revit utilises a native.rvt file type for storing BIM models. Normally, a building is designed using parametric 3D objects to create walls, floors, roofs, structure, windows, doors and other objects as required. Utilising parametric change technology, any change is automatically updated throughout the project, ensuring the design and documentation are coordinated and reliable. These parametric objects - 3D building objects or 2D drafting objects (such as surface patterns) - are called “families” and are saved in .rfa files, and imported into the RVT database as needed. A proper database system or warehousing would be necessary (Chong and Zin, 2011, Chong *et al.*, 2011, Chong and Phuah 2013, Chong *et al.*, 2013).

Families are grouped by Category and sub-divided by Type as shown in Fig. **10** below:

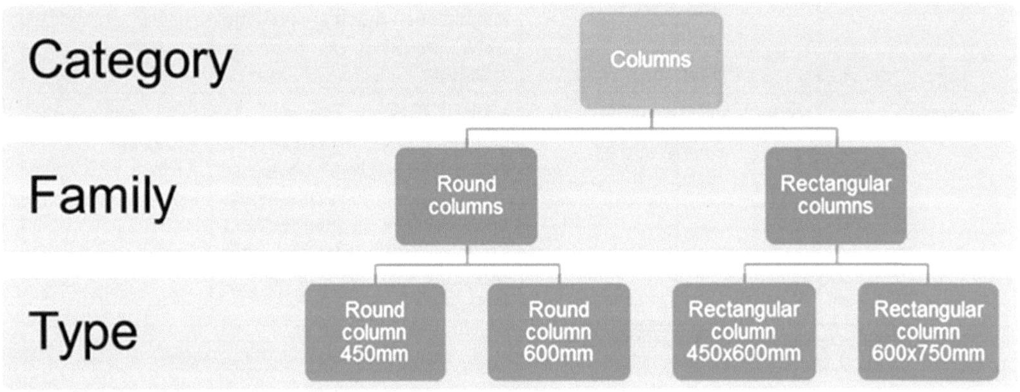

Figure 10: Autodesk revit - category, family and type.

Although many categories of families exist, they are classified into three groups:

1. System Families are typically the main elements of a building such as floors, walls, roof and stairs. Usually they have been produced by Autodesk, and are preloaded into a new Revit project. Although customization is limited, they can usually be sufficiently modified to suit project requirements. They perform in this manner so that the software can understand relationships and rules between them and other families.

2. Loadable Families - while System Families comprise the main components of a building, the remainder of the content in a building is comprised of loadable and in-place families. These can be generated, as parametric models with properties and dimensions, by the user from basic geometric shapes or downloaded from a wide range of sources on the internet. They include doors, windows, ceramic tiles, sanitary ware and soft furnishings. There are very few limits to what

can be created, including the parameters that describe the behaviour of the family. The user may change a selected family by modifying existing parameters such as length, weight or quantity. As such, a family describes the geometry which is controlled by parameters. For example, a window may be a Family. It may have types describing different sizes, features and installation requirements.

3. In-Place Families which are like retrievable families, but cannot be saved out of the project in which they are developed.

Fig. **11** illustrates the default family categories typically loaded into Revit. They are provided with the software and as described above, can be modified and added to so that most situations can be satisfied.

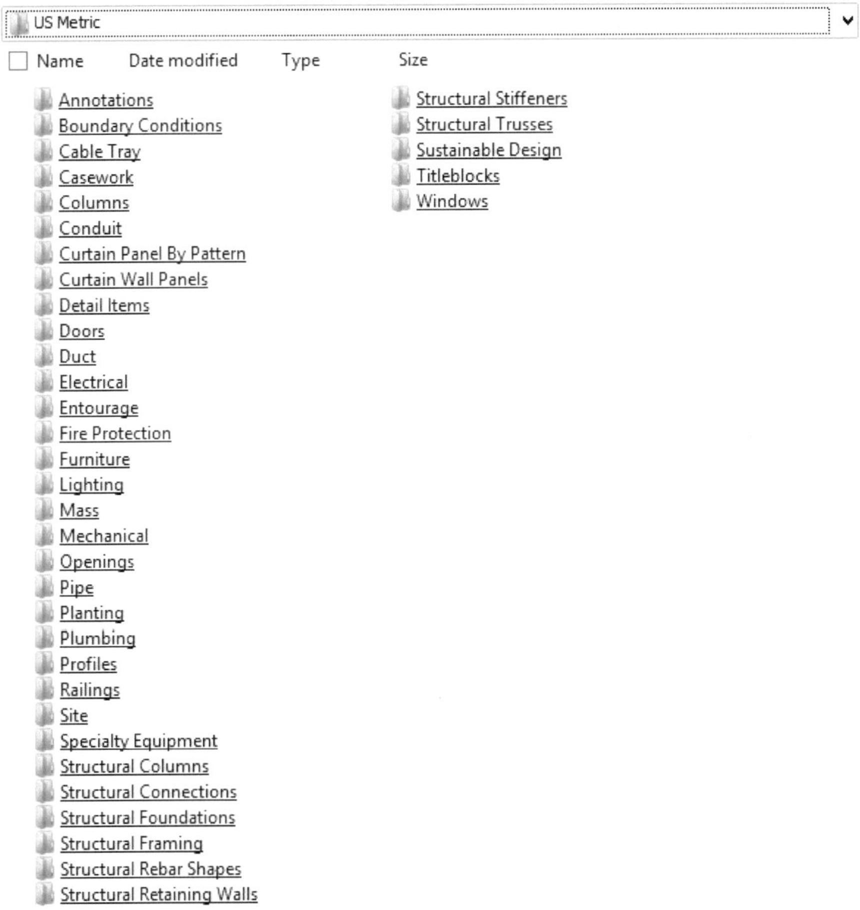

Figure 11: Revit 2014 default families. (Source: Developed for this book).

Fig. **12** shows the default door families inside the door category in Revit 2014:

Figure 12: Revit 2014 door families. (Source: Developed for this book).

Detailed information on System, Loadable and In- Place Families within Revit Architecture, Structure and MEP can be found on-line.

4.5.6. Rendering

Revit provides three native solutions to produce sophisticated renders and animations. These allow designers, to generate images that communicate the design intent more visibly to other project stakeholders. With Revit, a user has the option of using the Ray Trace Visual Style, the mental ray rendering engine or the Autodesk 360 Cloud Rendering service. The three rendering options have particular outputs with respective advantages and disadvantages. A number of third party rendering applications and plug-ins provide a further range of solutions.

With a moderately powerful desktop computer, Ray Trace is a real-time rendering environment. A 3D view in Revit generated by Ray Trace, simulates the path of light rays take as reflect from surfaces, mimicking real world experience. A particular benefit of Ray Trace is that it generates an interactive view. As it is renders, the user can continue to navigate around the building and the render image will adjust and re-render in real-time. As such, the output quality is less photo-realistic than the other two options, although this will improve as computing power increases. Hence, Ray Trace is useful for dynamic, interactive renderings where highest quality is not needed.

To render an image to a high quality is best achieved with a powerful desktop computer and time. This increases with the specification of glossy or transparent materials on objects. Mental ray provides many settings to adjust the quality of the final rendered image, including transparency, reflection, precision and indirect lighting and sky illumination settings. Use of the advanced settings often improves the quality of the output, but typically requires a longer time to generate, especially if several of the settings are increased simultaneously. Many settings may only add a few minutes to the rendering time, however, some may add several hours or sometimes days. The mental ray rendering engine can generate a high level of quality specified from a sophisticated toolset of advanced options, however, complex images can take a long time to render.

Rendering a complex image to a high quality can take a very long time and the computer performing the render cannot be used for other production work. One solution to this predicament is to use the Autodesk 360 Cloud service. In this arrangement, information from the user's Revit model is uploaded over the internet to Autodesk, where the rendering is performed on the Autodesk rendering engines in the Cloud. Other than the upload and download, the user's computer is not used and is available to carry out other work during the rendering process. To use this service, users must have an Autodesk subscription that is bought as an additional cost to the software. Typically, an image that would take Revit an hour to render will usually take less than 5 minutes using the Autodesk Cloud. Although the Cloud settings are very similar to the settings in Revit, the rendered images from the Cloud are not generated by mental ray, thus a Cloud rendered image may appear different to the locally rendered image.

With an adequate internet connection the Autodesk 360 Cloud rendering service provides an extremely fast solution and to a high standard, although without the highest levels achieved by mental ray. As with a rendered mental ray image, the render will look slightly different than the same scene rendered with ray trace.

4.5.7. Collaboration

Construction projects are typically collaborative ventures between a number of parties. Specialist collaboration software is becoming increasingly available. This concerns the digitalisation of the information flows between project participants, utilising software such as Aconex, Newforma or Asite. It overlaps with BIM collaboration software, including Autodesk tools (360), Bentley tools, VEO, Cadac and burgeoning list of others. Table **14** describes some of the functionality that may be provided by the collaborative software solutions:

Table 14: Areas of collaboration

Function	Description
Metadata	Metadata describe as the date the document was kept and the identity of the user who storing it.
Integration	Incorporate document management directly into other software,. Such integration is commonly accessible for office suites and e-mail or collaboration/groupware software.
Capture	It is about accepting and processing images of paper documents from scanners or multifunction printers. Total stations and point scanners/lasers.
Indexing	Indexing is designed to support retrieval. A rapid retrieval is important for the creation of an index topology.
Storage	Storage of the documents often includes database management. It should be arranged as per hierarchical storage management.
Retrieval	Flexible retrieval allows the user to locate partial search terms involving the document identifier and/or parts of the expected metadata. It will be supported certain retrieval systems, for example, Boolean.
"Distribution"	"A published document for distribution has to be in a format that cannot be easily altered. As a common practice in law regulated industries, an original master copy of the document is usually never used for distribution other than archiving. If a document is to be distributed electronically in a regulatory environment, then the equipment tasking the job has to be quality endorsed and validated".
Security	It is about certain access rights to the documents. Some document management systems would allow an administrator to grant access to documents based on type to only certain users.
Workflow	There are different types of workflow. "Manual workflow requires a user to access the document and choose whom to send it to. Rules-based workflow gives an administrator a right to create a rule to transfer the document.:. Dynamic rules allow flexibility in the workflow process".

Function	Description
"Collaboration"	Collaboration is about shared working environment. It allows multiple users to view and modify (or mark-up) a document at the same time at the working platform.
Reproduction	Document/image reproduction is key for building plans. It considers how plans will be scanned and scale will be retained when printed.
Email support	Registering and allocation of emails to project data.
File types	File type compatibility and limitations.

Source: Developed for this book.

Chapter 5: The Way Forward

CHAPTER 5

The Way Forward

Abstract: The searching for the end-to-end application and effectiveness of the practice of BIM is still in process. The authors believe that the barriers and issues faced by BIM, are tackled through the usage of cloud computing or the web. A cloud based framework acts as a central server where all the designs and engineering software are located. These data can then be accessed by personnel and used in a local server before being synced back with the central server. This makes it more efficient for users to coordinate using BIM. Apart from that, the focus of the industry has been on how to make a basic BIM work, as opposed to exploring what can be achieved with BIM. This is changing as an ever increasing array of specialised products and services evolved. This is beginning to change as more government agencies express an interest in developing BIM-based workflows that add efficiencies to design approval and regulatory compliance. Besides, availability of library data in BIM is required to make BIM work effectively. In conclusion, BIM is evolving organically across a range of dimensions and building lifecycle phases.

After reviewing and analysing BIM's current developments and philosophies, it is vital to know and forecast the future and trend of BIM from different aspects. It classifies into two perspectives, which is referring to academic and industry backgrounds of the authors. It would provide a more comprehensive view on the way forward of BIM.

5.1. ACADEMIC PERSPECTIVE

BIM has proven to be very effective and beneficial in many aspects in numerous sectors, but there are several barriers that must be overcome in order to facilitate greater adoption of BIM in the construction industry, such as transactional business process evolution, computability of digital design information and meaningful data interoperability (Bernstein and Pittman 2005). In reality, the searching for the end-to-end application and effectiveness of the practice of BIM is still in process.

The authors believe that to tackle the barriers and issues faced by BIM, is through the usage of cloud computing or the web. It was agreed that web portals would be necessary for full collaboration and sharing of data as required by BIM (Isikdag, 2012). This can be a cloud computing based BIM. For example, cloud computing based smart building models have recently been developed, as shown by Sawhney

John Rogers, Heap-Yih Chong, Christopher Preece, Chai Chai Lim and Himal Suranga Jayasena

and Maheswari (2013). In their proposed framework, a cloud based framework acts as a central server where all the designs and engineering software are located. These data can then be accessed by personnel and used in a local server before being synced back with the central server. This makes it more efficient for users to coordinate using BIM. Besides, the construction industry can benefit most from cloud computing through Software as a Service (SaaS) infrastructure and data virtualisation (Underwood and Isikdag, 2011). This is because through cloud computing, all the softwares and hardwares required can be made through the cloud with virtualisation, thus eliminating the need for companies to acquire software and physical hardware, resulting in extensive savings in cost. The cloud computing coupled with BIM could improve interoperability among BIM applications by sharing and accessing information across multiple platforms, yet security and legal aspects should be well taken care as some information can be sensitive and require confidentiality (Redmond *et al.*, 2012).

Apart from that, the management and contractual aspects of BIM are the main agendas for the growth and maturity of BIM. It is because existing procurement and project delivery approaches have remained relatively unchanged for many decades and they should be fine-tuned to suit the different characteristics and working environments as adopted in BIM. A workable legal framework is still at preliminary stage (Olatunji, 2011). Besides, Management for BIM (M-BIM) aims to coordinate and manage overall information, process and strategy for a whole project lifecycle, yet the implementation of M-BIM requires a thorough investigation in terms of the supply chain management, contract administration and overall project management.

5.2. PRACTITIONER/INDUSTRY PERSPECTIVE

At its most basic, the concept of BIM can be reduced to a database with a data structure such as IFC. Now, the base data that describes a building's form and functions are increasing, albeit relatively rarely, in this format, the opportunities and possibilities afforded are only now being ruminated on by the AECO sectors.

Until recently, the focus of the industry has been on how to make a basic BIM work, as opposed to exploring what can be achieved with BIM. This is changing as an ever increasing array of specialised products and services evolve. Particularly it is seen in the support sector where, for example, Autodesk have formed a strategic alliance with site equipment specialist, Topcon, and site equipment specialist Trimble have purchased Sketch-Up from Google and 5D BIM specialist Vico. This movement of BIM to site is being experienced in the industry, with clients at the leading edge regularly asking for support in this transition. Furthermore, increasing interest is being shown by client operators who are beginning to appreciate the value of accurate and detailed 'as-built' models for their operations and facilities management activities. Therefore, the BIM would work better in terms of its implementation on site and overall operations and facilities management in the near future.

Next, it is probable that the project phase transition will in turn drive the use of mobile devices across the industry. As processing power follows Moore's Law, powerful, mobile devices will enable designers and construction phase operatives to work effectively at the optimum location, be it site or studio. This will be enhanced by the increasing and symbiotic shift to cloud based storage and processing. Indeed, some software providers actively encourage users to export specialised processing tasks, such as Autodesk's cloud rendering system, or their Ecotect environmental analysis (and option generation) system. It is likely that, following history, the adoption of these services into a project team's workflow will experience difficulty; however, this time, the difficulties are more likely to be as a result of human resource limitations than technical.

Apart from that, although technology solutions are widely available, most firms still rely on traditional design and build processes. This is beginning to change as more government agencies express an interest in developing BIM-based workflows that add efficiencies to design approval and regulatory compliance. As the regulator and the largest client, this influence will permeate down to the rest of the industry and drive increased adoption at every level. This soft aspect is a significant area and will drive BIM forwards.

Finally, the availability of library data is increasing exponentially. This is coming from two sources. The first are the in-house generated objects or families that a design practice will create for a project. Digitised information is inherently suited to copy and pasting or re-using. As design practices increase their experience, their own libraries are increasing. Secondly, equipment and product manufacturers are increasingly making their product catalogues available as BIM ready objects or families. This ability to select pre-drawn objects from a library and simply drop them into a design is bringing enormous gains in productivity, efficiency and effectiveness. This is naturally facilitating the linking up with the extended supply chain, further driving improvement in a virtuous circle. It would be another important area of development and implementation of BIM in supply chain management.

In summary, BIM is evolving organically across a range of dimensions and building lifecycle phases. Amongst the leading practitioners, using a metaphor, BIM has emerged from a crawling phase and is beginning to walk with increasing speed and confidence.

5.3. ROADMAP FOR THE ADOPTION OF BIM

In order to be successful in the implementation of BIM the stakeholders must understand the followings:

Point 1: Role

Source - Changing roles of the clients, architects and contractors through BIM - Rizal Sebastian, TNO Built Environment and Geosciences, Delft, The Netherlands, Engineering, Construction and Architectural Management, Vol. 18 No. 2, 2011 pp. 176-187.

- An effective multidisciplinary working platform requires different roles for the project stakeholders; new contractual relationships; and new collaborative working environments. It will create new working platform where the related stakeholders to be involved at the early stage of the life cycle, namely, planning and design phases. The implications of the new strategy will bring different roles in the different project delivery method.

- The success implementation of the collaboration is also depending on the client's capability to use a collaborative tendering strategy.

- The implantation of BIM will provide an optimal cross-disciplinary and cross-phase collaboration creates new job opportunities.

Point 2: Processes/Standards

Source - Construction Management and Economics, 2013 - The advantages of information management through building information modelling by Peter Demian and Davis Walters - School of Civil and Building Engineering, Loughborough University, Loughborough, UK.

- "In order to gain the full benefits of BIM, these changes must be made gradually and within multiple collaborating organizations, with decisions regarding the implementation made on a project by project basis".
- "A broad category of process changes relates to the significant volume of training required. A relatively steep learning curve is associated with a switch to BIM technologies".
- "From a contractual standpoint, the increased collaboration between organizations employing BIM means an increased entwining of fortunes. Current contract terms do not allow for this collective responsibility, nor do current tools provide enough support for tracking and monitoring changes. The closer collaborative working facilitated by BIM also highlights the problem of interoperability".
- "Standards are common throughout the construction industry already but the implementation of BIM requires the development of new standards, particularly those specifically for construction".

Point 3: Interest

Source - BIM and the public interest by Richard Lorch - Building Research & Information (2012) 40(6), pp. 643-644.

- One significant question is where this information on buildings is to be held. Should it reside privately with clients and building owners, be shared amongst the supply side, be placed in the public domain (*e.g.*, within a city level model), or be given to a public interest organization? Is all the information in a digital model appropriate to be placed in the public domain? There may be legitimate reasons to redact certain kinds of information from public scrutiny which are commercially confidential or that might compromise the safety of a facility. There may also be a need to differentiate levels of access to information by different users - the general public, practitioners, local authorities, researchers, policy-makers - and to provide access to different levels of granularity.

Point 4: Challenges

Source - Building Information Modeling (BIM): Trends, Benefits, Risks, and Challenges for the AEC Industry - By SALMAN AZHAR, - Leadership and Management in Engineering JULY 2011

- The BIM adoption has been much slower than anticipated especially in the local construction industry. It is purely caused by technical and managerial factors.
- The technical factors can be explained as follows:
- "The requirement for a new collaborative construction process model to remove data interoperability issues",
- "The requirement that digital design data be assessable", and
- "The requirement for a new integrative approach or the purposeful interchange and incorporation of useful information among the building information model components".
- The managerial reasons are:

"At the moment, there is no clear guidelines on how to effectively adopt and use BIM. Numerous software vendors are working hard to catch up of the BIM evolution and have figured out certain quantitative aspects of it. It is important to normalize the implementation process.

Another important matter among the stakeholders is who should develop, operate or even own the building information models and the right way to share the costs involved.

Researchers and practitioners have to develop suitable solutions to overcome these challenges and other associated risks".

These are the issues that have to be overcome in order to move forward.

CHAPTER 6

Case Study 1: BIM in Malaysia

Abstract: Although it can be seen that pockets of BIM awareness exist in Malaysia, there is no comprehensive picture of the level of awareness or perceptions across the Malaysian construction industry. There are insufficient BIM competent people in Malaysia. There are grave concerns pertaining to the legal issues that may or may not be attributable to BIM, including protection of IPR and the implications for professional liability. The Malaysian Government may have policies intended to support BIM adoption but these are unknown to the industry. Overall, there is a strong belief that acquiring a BIM capability will provide a competitive advantage in the market place. It is recommended that the government should appoint a BIM champion and Task Force to promote BIM development in Malaysia

6.1. BIM MATURITY IN MALAYSIAN CONSTRUCTION INDUSTRY

The following sections discuss and review the status of BIM in Malaysian construction industry.

Awareness

BIM is described as bringing many benefits to the AECO industry (Autodesk, 2011, Baba, 2010, Becerik-Gerber and Rice, 2010, Bew and Underwood, 2009, CIB-W078, 2010, Eastman *et al.*, 2011, Lu and Li, 2011, Singapore-Gov, 2010, Underwood *et al.*, 2011, Young, 2009), and can deliver a range of product, operational and business benefits, as summarised by Coates, Arayici, Koskela *et al.*, (2010):

Short Term Gains

- Improvement in the quality, speed and cost of the services
- Automatic low-level corrections when changes are made to the design through the use of parametric relationship between objects
- Generate accurate and consistent 2D drawings throughout the design
- Visualizations to allow checking against design intent
- Discovering design errors before construction

John Rogers, Heap-Yih Chong, Christopher Preece, Chai Chai Lim and Himal Suranga Jayasena

Medium Term Gains

- Information sharing
- Greater flexibility to satisfy customers
- Better financial control
- Simultaneous work by multiple disciplines

In attempting to quantify the projected benefits of BIM, Morell (The-NBS, 2012b) suggests that a typical construction project experiences a waste level of approximately 25%. Of this 25%, he estimates that a fully functional BIM implementation would make a saving of approximately 50%. Thus, a well deployed BIM implementation would save approximately 12.5%, or one eighth, of project costs. Although Brewer (2010) cautions that BIM is not mature and that full implementations are not commonplace, the projected potential savings generated by BIM should be sufficient to attract industry awareness. Annual BIM surveys from the UK and New Zealand do suggest that there is a growing and significant awareness of BIM in their national AEC industries (The-NBS, 2012a, Masterspec, 2012).

BIM's use is proliferating around the world. Powerful drivers are converging from across the environmental landscape which may profoundly impact the market, the firm and the firm's operational processes. As is evidenced by government policy in countries such as UK, Singapore, USA, Australia *et al.*, the adoption of BIM is becoming mandatory. Other countries and regions, such as Hong Kong and parts of China have demonstrated expert competency in BIM and are beginning to provide empirical data to support the claims that BIM delivers significant improvements in time, cost and quality, and also against a much wider range of metrics.

BIM's adoption is advocated by authorities, governments and industry bodies. Research relating to the improvement initiatives of IBS (Fathi *et al.*, 2012a, CIB-TG57, 2010, CIB-TG74, 2010, CIB-W102, 2010), Lean (Sacks *et al.*, 2010b, Arayici *et al.*, 2010, Arayici *et al.*, 2011) and IPD (Matthews and Howell, 2005, Eastman *et al.*, 2011) all identify that the common and essential facilitating technology is BIM. Without the capability that BIM provides, little of the projected benefits of these initiatives will be realized.

Globally, BIM is beginning to demonstrate superior performance across the dimensions of time, cost and quality and the realized and purported benefits can be categorized into these groupings. As such, awareness of the benefits of BIM can be similarly categorized.

Despite Malaysia having an advanced IT industry, (including a development team for BIM software company Tekla) (Tekla, 2012), little academic research has been undertaken into BIM in the Malaysian context. At the time of writing, only two pieces of academic research addressed the topic. The findings of Tan's (2011) Bachelor's degree survey questionnaire study identifies that the level of awareness of BIM in the quantity surveying profession is low. Baba's (2010) Master's research solicited response, *via* questionnaire survey, from only twenty nine respondents of the architecture and engineering professions in the Kuala Lumpur region.

Therefore, it can be seen that pockets of BIM awareness exist in Malaysia, however there is no comprehensive picture of the level of awareness or perceptions across the Malaysian construction industry, and following Gu *et al.*, (2008) this represents the necessary first step in understanding the adoption of BIM within the Malaysian scenario.

Barriers to BIM Adoption

BIM is described in the literature as complex, difficult to implement and expensive (Eastman *et al.*, 2011, Lu and Li, 2011, Roper, 2012). Contrary to the clear benefits that BIM brings to a project are the difficulties faced in implementing BIM. Brewer (2010) articulates some of these difficulties and contrasts the chasm between the benefits suggested by the conceptual component of BIM and the difficulties of the applied reality. This concurs with Succar (2009) who coined the pejorative phrase 'BIM-Wash' to describe the dislocation between BIM potential and BIM reality. With recognition to the difficulties of BIM adoption, research has also been undertaken to facilitate uptake (Gu and London, 2010, Roper, 2012). In the Malaysian context, Teo identifies the expense, lack of suitably skilled human resource, and organisational and process difficulties as barriers to BIM adoption (Teo, 2012). Also in the Malaysian context, Baba (2010) identifies technical (interoperability), process, cost, legal, human resource skills as barriers and market demand.

Financial Considerations

BIM typically requires new software and regularly requires new or upgraded hardware to run the processing intensive software (Autodesk, 2011, Eastman *et al.*, 2011). There is a large technology component to a BIM implementation. Although concerns continue over the inoperability of some software and hardware platforms, this can be overcome by providing all players with the same software, or at the least, software from the same vendor. Any technical barriers to adoption can therefore be eliminated, however this repositions the problem as a financial issue, as such, there is a financial impact associated with adopting BIM.

The current USA price of Autodesk's BIM entry level software, Building Design Suite Premium, is US$6,825 (Autodesk, 2013). The current Malaysia price of Autodesk Building Design Suite Premium RM19,136. This equates to an exchange rate of US$1.0:RM3.20. Authorities on Purchasing Power Parity indices suggest that. *"money market exchange rates are unsuitable for converting costs to a common base as they reflect changes in the money market and not value"* (Best, 2010, p. 117). A crude measure that finds acceptance with authorities for heuristic comparison (Best, 2010, Meikle, 2011, Langston and Best, 2012, Clements *et al.*, 2012) is The Economist's Big Mac Index (The-Economist, 2013). This suggests an exchange rate of RM1.82:US$ as being more equitable, and would result in a Malaysia price for Autodesk Building Design Suite Premium of RM12,421 which represents a 41% reduction. Further examination of PPP and pricing structures is outside the scope of this research, however, it potentially represents an area for further research.

These costs indicated above are only for the purchase of basic BIM software and do not include costs for training and downtime as the company internalises new working processes. Adoption of BIM is a major financial investment.

Process Changes

The MacLeamy curve provided in figure identifies the fundamental nature of the change to a BIM adopting organisation's operational processes. The entire design process is brought to the front of the project and executed earlier and to a higher level of completion than traditionally so. This concurs with the Bews-Richards BIM Maturity Model, shown in figure, that illustrates the successive levels of sophistication that are needed for an organisation to adopt increasingly sophisticated levels of BIM.

With regards the technology platforms that BIM rests upon, the majority of these are provided by proprietary software development companies, such as Autodesk, Bentley, Graphisoft or Nemetschek (Masterspec, 2012, The-NBS, 2012a), which, despite protestations to the contrary, are considered by expert users to be imperfectly interoperable (AEC-(UK), 2012b).

Human Resource

Adopting BIM requires fundamental process change within an organisation and with it, a complementary change in the skill sets of the human resource pool. Following a Bews-Richards or AIA definition of minimum BIM, this will also entail concurrent capability increases along the project supply chain, including developers, other designers, contractors, approvals authorities, all having personnel with the competency to adopt BIM.

Legal Factors

Level 2 BIM requires collaborative working relationships between design and project team members (American-Institute-of-Architects, 2012b, American-Institute-of-Architects, 2012c, American-Institute-of-Architects, 2012d). The UK's Construction-Industry-Council (2013, p.v) states as a key objective of its BIM Protocol, *"In light of industry concerns in respect of IPR and the increased collaboration involved in a BIM project, clause 6 of the Protocol clearly sets out the IPR provisions required to enable the Models to be used as intended and to protect the rights of the Project Team Members against infringement".*

However, it remains that 95% of works carried out in Malaysia is executed with traditional contract types (CIDB, 2012b). This situation merits close attention; however, it requires specialist legal knowledge which is outside the training of this researcher. A thorough analysis is therefore delimited for this research, although it will remain a variable to be considered, although imperfectly understood. This represents a weakness in this research, and denotes an opportunity for further research.

Professional Support

The boundaries of a Malaysian consultant firm are mandated and constrained by Malaysian Government Charter. Contrary to the geographical boundary-less nature of electronic data interchange and BIM, nations still maintain fixed geographical boundaries and distinct cultural and political identities, including the management of industries, trades and professions within their borders. Although there are movements within the ASEAN region to dismantle some of these constructed knowledge and commercial barriers, with the exception of mega-projects, the vast majority of the indigenous market is serviced by local companies, operating under local mandates, to local guidelines with local staff. As such, BIM needs to be tailored for adoption in the Malaysian consultant context, and in so doing, creates potential for nationally idiosyncratic barriers to the adoption of BIM as well as common barriers found across all countries. In so doing, this creates an indigenous contextual component. Allied to the bespoke nature of consultant service, it is necessary to identify any peculiarities of the Malaysian consultant instance and what general or specific barriers will be found.

As evidenced by (AEC-(UK), 2012c, Autodesk, 2011, Bew and Underwood, 2009, Construction-Project-Information-Committee, 2012, Eastman *et al.*, 2011, Young, 2009) successful BIM implementations typically receive the support of knowledgeable persons or consultants that have specialist expertise

Governmental Supports

The Malaysian Government has manifold interest and influence upon the construction industry. Firstly, it is the regulatory body that sets the boundaries and controls the behaviour of professional designers. Secondly, it is the approvals body for land and building applications and the quality control body for establishing building codes. Thirdly, it sets the strategic direction of the industry, directly through plans such as the Construction Industry Master Plan, and indirectly through economic policy. Fourthly, it is the industry's largest client. Finally, it is the responsible body for the calibre of human resource provided through its educational system.

The Malaysian Government's national intent is articulated in the 10th Malaysia Plan (The-Economic-Planning-Unit, 2010) and calls for large scale infrastructure spending. This is allied to Vision 2020, the national goal of achieving developed nation status within eight years. Similarly, this requires considerable infrastructure investment and correspondingly advanced professional services delivery.

Promotion

The principal objectives of the Malaysian Government vis-a-vis the direction of the construction industry are contained within the Construction Industry Master Plan (CIDB, 2006, p. 10).

Generally this calls for an integrated value chain, improved quality, productivity, and health and safety, human resource capability building, and a view to exporting professional skills overseas. Specifically it extols the benefits of IT as being the most appropriate enabling tool, for which BIM can be considered as the most powerful manifestation. Consistent with these stated objectives would be evidence that the Malaysian government is proactively promoting BIM to the AECO sector with a correspondingly high level of awareness within the sector.

Financial Support

The Malaysian construction industry has poor profitability and high firm failure rates (Ibrahim *et al.*, 2010a). This is exacerbated by the relatively modest Gross Fixed Investment, shown in Section 2.6.1 and Tables **2-7**, which is further exacerbated by the global economic downturn. The Construction Industry Master Plan (CIDB, 2006, p. 8) prescribes a '*Progressive industry employing highly skilled workers using modern techniques and technology*' and an '*Innovative industry - that benefits from structured application of R&D initiatives*' Given these drivers, and the cost impediment described above, it would be feasible that the Malaysian Government provide financial subsidy. This would follow the example of such countries as Singapore, whose Government subsidises its construction sector to adopt BIM (Singapore-Gov, 2013, Singapore-Gov, 2011, buildingSMART-Australasia, 2012).

"The CORENET e-Plan Check system is a program that aims to encourage the construction industry in Singapore to use BIM. The program is fully funded by the government and is designed to move the construction industry from 2D design to building information models that are data rich and are used throughout the life cycle of the building from design through to construction and demolition" (buildingSMART-Australasia, 2012, p. 47).

Professional Bodies

The charters of the professional construction bodies each aim to enhance the knowledge, technical competency and capability of their members. As such, it would be expected that campaigns to promote BIM would be undertaken or advanced by the professional bodies, and this has been the case with ACEM and

PAM (ACEM, 2012a, PAM, 2012). Awareness driven by these bodies would be expected to be apparent in the bodies' members.

Compulsory BIM

Since 2009 it has become mandatory that all public sector projects exceeding RM50m must be validated through Value Management techniques. With regards to IBS, the Government has mandated that all public-sector projects must attain no less than 70% IBS content under the Treasury Circular SPP 07/2008. As a client, the Malaysian Government, accounts for 24% of total construction projects, by value, in Malaysia, as shown in Tables **2-10**. As such, should BIM be considered a strategically important step in the Malaysian government's industry sector development plans for the AECO industry, it would be reasonable to observe tangible government activity in its procurement policy with regards to compulsory BIM capability. This would follow strategies imposed by countries such as the UK and Singapore.

6.2. FOCUS GROUP INTERVIEW

As there is a positive indication that Malaysia is adopting BIM. The involvement of local experts and their knowledge related to BIM would be an added value to the book. Therefore, focus group interview was adopted in this case study.

A focus group is *"a research technique that collects data through group interaction on a topic determined by the researcher"* (Morgan, 1997, p. 6). Participants are encouraged to express their different points of view and the aim is not to build consensus nor to determine the strength of their opinions (Stewart *et al.*, 2007), as, *"Crucially, focus groups are distinguished from the broader category of group interviews by 'the explicit use of group interaction' as research data"* (Kitzinger, 1994, p. 103). In so doing, the focus group examines, *"...not only what people think, but how they think and why they think that way"* (Kitzinger, 1995, p. 300).

In addition, focus groups allow the researcher to *"obtain deeper levels of meaning, make important connections, and identify subtle nuances in expression and meaning"* (Stewart *et al.*, 2007, p. 16). Focus groups 'offer the researcher the opportunity to observe directly the construction of meaning in a social context *via* interactions of group. Table **15** shows the strengths and weaknesses of focus groups.

Table 15: Strengths and weaknesses of focus groups

Strengths		Weaknesses	
1.	allows observation of group interaction in controlled setting	1.	relies entirely on group interaction
2.	good for less structured themes/topics	2.	not naturalistic - created by a researcher
3.	very helpful for exploratory research	3.	may not work for all topics
4.	allows the researcher to produce concentrated data on a precise topic	4.	participants may steer discussion in directions not relevant to the research
5.	quick and easy to conduct	5.	some participants may dominate discussion
6.	participants can often exert more influence over the discussion than in one-to-one interviews, so less of a power imbalance	6.	raises ethical issues about confidentiality - everything said is shared with other participants
7.	can provide rich and in-depth data	7.	may not be suitable for people lacking confidence or relatively inarticulate
8.	is a flexible method	8.	may prove difficult to recruit participants

Source: Adapted for this research from (Deem, 1997).

Following consideration of the weaknesses, it was decided that the weaknesses could be largely mitigated. The advantages of the focus group justified its use as a primary data collection tool for the qualitative component of this research.

6.2.1. Selection and Composition of the Focus Group

There is no prescribed number of focus groups sessions required for a research project. Different authorities suggest different numbers and there is little consensus on what is most appropriate. The number of groups is contingent upon the purpose of the research, the background information needed, the nature of the focus group, and the success of the proceeding focus group, if any. Due to time and resource constraints, the researcher decided to engage only one focus group with the consideration that there will be a complementary web survey later.

The participant group size similarly varies according to authority and purpose; (Zikmund, 2003a, p. 117) recommends a range of six to ten people per focus group. Kitzinger and Barbour (1999) recommend four to ten. The size must be large enough to generate diverse viewpoint for generation of in-depth information but not too large for adequate participation by each group member (Zikmund *et al.,* 2010). Guided by these authorities, the researcher recruited ten focus group participants as this number falls within the range recommended by most authorities.

The judicious selection and recruitment of participants for a focus group is essential (Kitzinger and Barbour, 1999). Although some researchers favour an homogenous group, such a group with similar lifestyles, job classification and experiences will not have too many arguments and different viewpoints arising from the diverse background of the participants (Zikmund *et al.,* 2010). Kitzinger (1995) also supports diverse groups owing to the advantage bestowed of exploring different perspectives on the research problem. Stewart *et al.,* (2007) compare the relative merits of homogenous and heterogeneous groups, noting that too heterogeneous a group may stultify interaction and expression, whereas too homogenous a group may produce myopic output. Instead, *"In focus group interviewing the key to success is making the group dynamic work in service of the goals and objectives of the research"* (Stewart *et al.*, 2007, p. 20), and advise *"appropriate blending or selection of participants".*

The existing literature pertaining to BIM in Malaysia indicated a low level of adoption. Furthermore, it was evident that there was a low level of understanding of the concept of BIM, and a low level of appreciation as to the impact to business and operational processes, and human resource implications of the decision to adopt BIM. This coincided with the researcher's professional experience of the reality of BIM implementation at corporate and project levels in Malaysia. Consequently, it was necessary to select participants that would, individually or collectively, provide a comprehensive and deep depiction of:

- strategic corporate leadership and corporate change of a firm - specifically business process change and human resource competency
- the industrial landscape and marketplace
- educational and training environment
- sensitivity to government policy
- experience of collaboration between design and project team members
- BIM technological excellence

The invited participants were stratified according to business and organizations covering the consultant firms, government construction bodies, professional institutions, developers and allied specialists.

Using the personal and business contacts, the following experts were contacted:

- Four BIM practitioners who used the BIM in their firms
- The ACEM was also requested to nominate a participant.
- The CIDB, the government agency responsible for construction, were asked to nominate one participant.
- The PAM was also requested to nominate a participant.
- The Chartered Institute of Builders (CIOB) was asked to nominate one participant.
- A Malaysian University providing BIM education modules was asked to nominate an academic familiar with BIM.
- BIM consultants - (one MEP Engineer, one Strategist)
- Two Architects with BIM experience
- Software vendor/developer

6.2.2. Planning the Focus Group Meeting

A meeting agenda was prepared with two objectives; firstly, to guide the meeting and facilitate group discussion, and secondly, to focus the discussion on the key research issues related to the research propositions.

6.2.3. Conduct of the Focus Group

Authorities recommend a facilitator be appointed to conduct and co-ordinate proceedings, to prevent the group from overly digressing and help retain focus on the discussion topics. Authorities recommend that group facilitators should stay at the process level and not be involved with the content (Krueger and Casey, 2009). For the purpose of this focus group, the researcher undertook the role of facilitator.

The discussion was free flowing and the participants were encouraged to express their candid views and comments on a range of issues related to the topic. The resulting group interaction provided valuable data and information on the research topics. Any pertinent points made by the focus group members that fell outside the scope of the discussion guide were treated as 'unplanned data'. The meeting was initially audio recorded; however, as some of the issues were commercially sensitive, at the request of participants, the audio recording was stopped and replaced by detailed written notes.

The focus group meeting was convened on 28th January 2013 in the BIM training centre of a local BIM consultancy. The ninety minute meeting was within the typical duration of one to three hours (Krueger and Casey, 2009).

6.3. ANALYSIS OF FOCUS GROUP INTERVIEW

This section presents the analysis of the data obtained from the focus group meeting.

Awareness

The concept of BIM in the engineering community is not new; it has been around since the 1990s when 3D modelling was employed in the Oil and Gas industry of which Malaysia has significant interests. Sophisticated 3D analysis, including clash detection, has been used in the structural steel sector for many years. Large and medium size firms already have exposure to at least some sort of 3D modelling. There is much industry confusion over what BIM is, with most engineers mistaking 3D CAD modelling for BIM. This misunderstanding is causing some engineers to perceive BIM as just another fad that will disappear. Overall, there is a growing awareness as evidenced by increasing numbers of seminars, conferences, workshops, publications, journals and media on BIM.

Design Quality

The 3D modelling and sophisticated analysis capabilities were considered to improve overall design quality; generally the final designs were more accurate and with less errors. The ability to visualise complex concepts was valuable, even for experienced engineers. The technical demands of the parametric 3D modelling process were considered as a constraint on creativity by the older generations but less constraining to the younger generations who grew up with IT. Owing to construction projects being prototype in nature, there is an inherent inability to prove design quality improvement as there are no control projects to compare against. There are no metrics to assess quantifiable improvements. Design quality indicators may have some utility but has not been tested anywhere.

Project Timescales

Clash detection was considered as a valuable tool for shortening overall project timescales by reducing site complications. The design phase was considered as lengthening owing to the added work burden of modelling in 3D, however, this was considered as a temporary condition that improves as familiarity and capability with 3D and BIM improves. The engineers that had mature 3D capability considered that office/studio productivity was greatly improved. Similar to design quality, the lack of a control project with managed variables dictates that assessments of project timescale improvements can only be subjective. Improvements in engineering firms' productivity are more objective and after an initial reduction there is a long term productivity improvement resulting from 3D modelling.

Project Cost

As with quality and time metrics, the lack of control projects prevents objective assessment. The speculation was that improved design accuracy, especially clash detection, would reduce variation orders and claims. There was consternation that some clients were beginning to request BIM capability of consultants yet were unwilling to pay higher fees. It was felt that higher fees were justified on the basis of improved and extended services being provided - value added - and the cost incurred in moving to a 3D and BIM environment. It was noted that the few, large developers that had achieved some modest BIM success were paying enhanced fees to consultants to support their adoption of BIM. These were considered exceptions but notably they were the organisations achieving dividends from BIM use. A small increase in consultants' fees as a percentage of overall project costs were considered trivial and frustration was noted that most clients were short-sighted and not attuned to benefits BIM could provide them.

BIM Maturity

The minimum BIM requirement of discipline models being shared and integrated with other disciplines were considered to be very rare in Malaysia with probably only a handful of authentic examples. The authentic examples were described as being undertaken by a few large and sophisticated developers with design and build capability. Most 'BIM' projects were, in reality, only 3D modelled projects. Some developers were beginning to pre-qualify consultants for BIM capability and that there was an emerging trend towards this, although it is still very small and at the lowest level. Estimations of projects using BIM is a maximum of one percent. One project, Wisma Shell, was cited as a successful example of Malaysian BIM use in which a LEED Gold certified 800,000 sq ft five storey office block was completed in 11 months [focus group participants had been project consultants].

The Cost of Adopting BIM

Unequal purchasing power parity puts Malaysia at a disadvantage with developed nations when buying the major vendors' software - the software generally costs the same in developed nations as it does in Malaysia, when compared to GDP/capita this equates to 2-4 times the cost paid in UK, USA or other developed nations. Specialist structural engineering software is particularly expensive. Generally, the software costs were considered a necessary evil and rarely was good value considered to be obtained from vendors. The trend in software costs was considered to be reducing and more competition was entering the market and driving down cost. Entry level software, such as basic CAD packages, and some 3D software, such as SketchUp were very cheap or free, and although not providing genuine BIM capability, they were useful in providing a BIM precursor capability that reduced the number of genuine/full BIM software licenses that were needed. Generally, the software costs were not now considered a significant barrier to entry. Hardware costs were not considered any barrier to entry and that the necessary processing power to operate genuine BIM software was becoming ubiquitous. It was also noted that the cloud processing services, for rendering and environmental analysis, offered by Autodesk, further reduced the need for expensive hardware. Software as a service (SaaS) is becoming a viable solution.

The largest cost of adopting BIM was described as the 'down time' required for individual and organisational learning. The cycle for individuals or teams adopting BIM (to a minimum BIM level where individual discipline models can be shared with design team partners) was described as taking a minimum 3-6 months before achieving an output comparable to what was achieved pre-BIM. Most of this learning time would be lost productivity. Organisational learning was suggested as taking a minimum 9-12 months. This was considered to represent a considerable investment for a firm and thus can cause serious financial difficulties to a firm. The failure rate of BIM adoption is suggested as 70% failure and this was significantly apportioned to the cost of downtime.

Furthermore, an additional cost of employing a BIM technician over a CAD technician was identified. BIM technicians generally need to be of a higher calibre and hence generally attract higher salaries regardless of whether they are using BIM or not.

Processes

The engineering community are experienced in working in procedurally driven environments. The new demands of 3D modelling have levels of support from BIM standards and the like, and that generally, these processes can be followed; however, there is inconsistency and lack of a common platform, noted as being ironic in a movement condoning collaboration and standardisation. Significant blame was placed upon the software vendors for developing vendor orientated solutions and that interoperability was still poor. The software was also described as ‘cumbersome’ and ‘clunky’ although iterative improvements were noted.

Generally, the complexities and idiosyncrasies of the overall construction process were deemed as a limitation and that poor understanding by many of the overall process hampered the introduction of BIM. A lack of understanding of BIM by adopters and a lack of a focused purpose for adopting BIM was identified as further impediments; successful BIM projects were considered as having a clear purpose for each BIM model. A lack of understanding of how BIM impacted the design and construction process was agreed as concerning and worthy of continued regard.

The Malaysian design process was described as being one where typically the engineering consultant provides single-line isometric or concept drawings and the trade or sub-contractors provide the detail that is then verified and adopted by the engineer. As the detail drawings are produced post-tender, it leaves little time for advanced analysis.

People

Considerable problems were identified regarding calibre, capability and quantity of human resources. Local graduates were lacking in BIM training by universities, a situation acknowledged by the university representatives. Correspondingly, the universities identified that the professional bodies were not accrediting BIM modules and without the accreditation there was little impetus or justification to include BIM modules in their syllabuses.

Considerable investment of time and money were required to teach students 3D modelling skills, however, without complimentary knowledge of the overall design and construction process, pre-BIM and with BIM, 3D modelling capability alone was insufficient to produce a valuable BIM modeller. People with a combination of 3D experience and real design studio or site experience are very rare, consequently, a serious skills shortage existed and firms were now seeking overseas staff to bridge this shortfall.

Key staff loss - through natural wastage and more ominously, poaching of trained staff by competitors - were identified as threats to project success. BIM was identified as placing considerable responsibility and power into the hands of often junior staff. The loss of one staff member could jeopardise a BIM project.

Legal

It was agreed that the contractual and legal relationship implications of adopting BIM (as a shared environment) were considered to be poorly understood and represented a notable risk to consultant firms and all design team participants. Once a firm had committed its design efforts to a BIM model that was then made available to a client, there was greatly reduced potential for ensuring client payment - traditionally the firm would release hard copy 2D drawings and only release electronic CAD files once payment had been received. It was agreed that the legal aspects of BIM required urgent investigation and action.

Professional Advice

Consistent with the human resource skills shortage, rapid technological changes and deficiencies in processes, there is a subsequent shortage of capable professional advice to guide adopters. There was a notable lack of contextualisation for the Malaysian situation, with particular note of the complex submissions process (including planning approval, building code approval and occupancy approval).

Software vendors were urged to provide more support.

Government Promotion

There was general consensus that the Government is organising awareness building activities. Singapore and Hong Kong are already far ahead with BIM adoption, however they actually adopted the concepts from Malaysia. There is a very large project going on which the Malaysian Government is a part of - they are using it as a case study. The Government is looking at a policy for a new economy and BIM is very much in the heart of it. CIDB is trying hard to push adoption of BIM.

There is a great interest to push BIM to the Malaysian Government at the industry level. BIM must be adopted "across the board" but the Government is very confused as to what BIM is and how it should be driven. The Government needs to play a role but what kind of role needs to be defined.

Government Financial Support & Subsidy

There was unanimous agreement that there is not enough support for BIM by the Government. Malaysian industries do not get many handouts from the Government, unlike Singapore. The large firms do not depend on the Government for any help.

Building plan approval is very tedious and designs keep changing - this hinders BIM use as model re-working is expensive. Furthermore, Government departments are far from being able to accept 3D submissions as Singapore BCA are now able to.

Singapore and Hong Kong are very easy countries to manage - Most governments try to be neutral - Malaysia is comparable to most European nations. Need for BIM in other countries is due to wider industry realizing the value.

Professional Bodies Encouragement

PAM and ACEM provide BIM awareness workshops and roadshows. Generally, the associations were considered to make a good effort but were hampered by meagre resources.

Mandatory BIM use on Public Projects

The Malaysian Government was urged to make BIM use compulsory as this would drive industry wide adoption. Consensus was reached that infrastructure is very important to national development and would be well served by mandating of BIM. Some government organizations are looking at the value in adopting BIM - Representatives from the Quantity Surveying bodies should push the government to adopt BIM.

The Government introduced the IBS agenda in 2003 [the policy intent is that 80% of public project work is prefabricated on projects exceeding RM50m], however, the AEC industry is still struggling to adopt IBS and this has caused a reluctance to impose BIM.

In Singapore the government has already made it a policy but Malaysia has no champion to impose pan-industries to provide the necessary platform, momentum and critical mass to ensure BIM adoption. In Malaysia the Civil society is much more dynamic than in Singapore - this allows a more active, and less compliant, industry; it is harder to enforce new regulations than in Singapore. The US and UK are already beginning to mandate the use of BIM

Market Drivers

There is much confusion over what BIM is, amongst Government, educators, industry bodies, vendors, specialists and users. There are many different perspectives, motives, experiences, needs and expectations. It is usually 3D CAD that is talked about. The engineering side is already aware but not sure if they actually know what BIM is. Is the demand actually for 3D modelling which is not BIM? If so, it is difficult to respond to market demands. As such, the demand for BIM is unclear because the understanding of BIM is confused. Needs more surveillance of the market to understand what is being demanded.

There is a dependency on Architects, as lead designer, to lead the adoption. It is necessary for Architects to start the industrial uptake of BIM. Engineers will not start it off.

Demand pull, capability push - there is a duality driving market; demand side and supply side. Some firms are beginning to offer clients value added services made available by BIM tools - especially sophisticated visualisations and walk-throughs - but not many clients are yet appreciating the value or benefits. Early forays into value-add BIM services are receiving a mixed response. At the market level, it is mainly the big projects that are adopting BIM.

Government, as the largest client, can be pivotal to adoption. Government should make it compulsory then the demand for BIM will increase.

Outsourcing

A few companies are offering out-sourced services, including out-sourced to overseas as there is insufficient capability in Malaysian market. There are large concerns over maintaining control and quality. Similarly, there are large concerns regarding loss of knowledge and IP. It is the nature of Engineering firms to do it themselves - it is what they do. Some tendency to in-sourcing, bringing in experts/resource from overseas and grow competency from that.

Adopt in Near Future

The use of BIM will definitely not happen within the next few years although it is important. Majority of engineers are taking an "on hold attitude" or a "wait and see" approach. Sunway has invested a lot of money and time in BIM software. [Strategic] partnerships are needed for the adoption of BIM. An industry wide option is needed. BIM adoption will happen but not so soon. BIM in Malaysia will be in the market in the next 5 years. BIM is for the future. BIM will be adopted in time but not so soon. Culture within the Malaysian industry is still too traditional - little or no CM or IPD, so the collaborative culture is not present.

Competitive Advantage

Some developers/contactors are paying enhanced fees to the Consultants to encourage BIM adoption. In Singapore the contractors are the ones who have adopted it - they have seen the benefit/value; growth of BIM in other countries is due to the wider AECO industries realizing the value, not isolated Engineering companies.

BIM provides "differentiation" but clients must understand why BIM is important then only will the overall industry change. The industry needs to adopt it as a whole with a teamwork approach.

6.4. CONCLUDING REMARKS

BIM is already providing a competitive advantage in the niche markets of sophisticated developers. Whether this becomes generalisable to the wider industry depends on BIM proving demonstrable benefit for clients, the clients sufficiently valuing the benefits for the supply chain to justify adoption. Should these conditions occur, the intention is to adopt BIM.

Some of the highlights of the findings are:

- There is an existing awareness of BIM supported by a level of understanding to make the complex phenomenon comprehensible. The industry has historic 3D capability and from this perspective appreciates that BIM has the potential to radically improve project performance across the dimensions of time, cost and quality.

- Currently there are considerable barriers to the successful adoption of BIM. These include the inherent difficulties of the organisational and operational change required by BIM. There is a skills shortage; there are insufficient BIM competent people in Malaysia. There are grave concerns pertaining to the legal issues that may or may not be attributable BIM, including protection of IPR and the implications for professional liability. The skills shortage extended to the professional guidance subsector supporting BIM adoption. Less significant was the cost of adopting BIM, although it was still a barrier. The cost of BIM software in Malaysia is high in comparison to fully developed nations.

- The Malaysian Government may have policies intended to support BIM adoption but these are unknown to the industry. The industry considers that the Government should be more dynamic and introduce policies that tangibly assist BIM adoption. Financial support for BIM adoption by firms, as in Singapore, and economic assistance to the professional bodies that guide the industry were recommended measures. The Government as a client mandating BIM, as in the UK, was also recommended to bring direction and critical-mass to the market.

- There is a strong belief that acquiring a BIM capability will provide a competitive advantage in the market place. There is strong intent to acquire this capability within the next two to five years. Either developing in-house capability or out-sourcing the BIM component of a project are viable adoption strategies.

Recommendations:

- The Government appoint a BIM champion to promote BIM development in Malaysia.

- The BIM champion should initiate an industry Task Force to chart a strategic direction for BIM in Malaysia. The Task Force should include knowledgeable senior figures from all Malaysian AEC stakeholder groups. The Task Force should also include renowned international BIM experts to ensure the inclusion of and alignment with world best practice.

- The Task Force would align BIM direction with IBS and VM initiatives to leverage the symbiotic relationship between them.

- The agency would take a proactive role in identifying and resolving BIM related issues. Close cooperation with the proposed Construction Court that solely adjudicates on construction contract disputes (Malaysian-Bar-Council, 2013) would enable Malaysia to be at the vanguard of construction law and lead any necessary BIM initiatives.

- The agency would introduce a web based portal to act as the authoritative source of information pertaining to Malaysian BIM. This would come to include BIM standards and the like.

- The agency's portal would continually harvest BIM information from users so that data was made available to researchers and students. This would support further and better quality research.

Chapter 6: Case Study 2: BIM in Sri Lanka

CHAPTER 7

Case Study 2: BIM in Sri Lanka

Abstract: Building Information Modelling (BIM) is not yet adopted in Sri Lanka. Improving the BIM awareness within the industry and beyond is a significant challenge in a BIM infant industry. BIMLab Network research group centred at the University of Moratuwa was the only active organization which has taken steps to increase the BIM awareness in the country at the time of writing. Besides, a wide variety of BIM software is made available by software vendors. Selection of software applications is primarily affected by the features of the software and the affordability. However, due to need of efficient information sharing and interoperability, the BIM Software Environment (BIMSE) would first affect the selection of software; or the selection of team members will depend on what software they use. Therefore, the decision of BIMSE becomes critical. Active involvement of Information and Communication Technology Agency (ICTA) of Sri Lanka would significantly change the context so that preferred BIMSE would shift to HSE. However, BIM is not in current ICTA agenda. An effort to convince ICTA of the benefit the country could receive by implementing BIM would become a significant leap in BIM adoption in the country.

Building Information Modelling (BIM) is not yet adopted in Sri Lanka. The case study below was reported and discussed based on observations and certain analysis of literature.

Being a developing country (IMF, 2012) and an industry with comparatively lesser standardized systems than those of developed countries, it should look for suitable adoption strategies to get the best advantage of BIM. Strategies and roadmaps being identified and adopted in developed economies with considerable levels of industry and technological maturity may not be readily adoptable in the Sri Lankan context. On the other hand, lack of maturity is also a blessing in disguise that it allows greater flexibility in selection of options for BIM adoption since there is no significant investment in technology to go waste by the selection of any of the possible options. However, an investor in BIM in such a context bears greater risk of making the wrong choice. Therefore, an industry with little or no BIM maturity should carefully study all possible options and select the best in terms of efficiency and effectiveness. This chapter presents a discussion of BIM maturity in the Sri Lankan construction industry and a review of a suitable software environment for it. The focus is on specific aspects related to Sri Lankan context, thus the content shall only complement the knowledge presented in previous chapters.

John Rogers, Heap-Yih Chong, Christopher Preece, Chai Chai Lim and Himal Suranga Jayasena

It should be noted that there were no real life BIM implementations in Sri Lanka to observe or to study their practical aspects. Thus, what is presented here is a theoretical review and a philosophical construct, by combining current knowledge with knowledge generated through the research conducted in Sri Lanka, and personal experience and knowledge gathered through industry interaction.

7.1. BIM MATURITY IN SRI LANKAN CONSTRUCTION INDUSTRY

The concepts of BIM maturity has been introduced and discussed earlier in this book. Even though BIM got a momentum in many parts of the world within less than a decade, the concept has been talked about even in early 70's (see Eastman *et al.*, 1974). However, it took a long time to become a reality that only in late 80's a software product became available for users known as ArchiCAD by Graphisoft (Laiserin, 2003). At time it was not known as BIM; Graphisoft called it Virtual Building, yet was the same concept known as BIM today. Keeping with this time line, the earliest BIM technology to come to Sri Lanka was ArchiCAD.

The 90's was an era which was quite conducive in terms of capacity building in computer application skills; especially among youth in the country. Developing computer skills soon after completing secondary education became a trend. Many private institutes started to offer training in various software applications and programming tools. While AutoCAD remained most popular among those who were interested in building and construction tools, a significant enthusiasm was also found to exist for ArchiCAD. The primary reason one would not like ArchiCAD was the lack of construction knowledge.

The trend was further supported by lack of legal endeavour to prevent software piracy. For the same reason, the trend did not become a commercial reality since these software became too expensive for local firms to obtain to be utilized for real life projects. The trend gradually, slightly diminished but still remains significant. The Intellectual Property Rights (IPR) Act was passed in 2003 to inter alia as an effort to tackle software piracy. After a decade, the software piracy rate still remains above 80% in Sri Lanka (2013). This supports unverified information being available that BIM capable software such as ArchiCAD and Revit are used unofficially in isolation, especially for small projects like houses. Due to the sensitive nature, this information is difficult be officially verified. In addition, most software vendors offer student licences and online tutorials (Autodesk, n.d.-b,

Graphisoft, n.d.-b, Bentley Systems, n.d., Vico Software, n.d., Exactal, n.d., Tekla, n.d.); many students in building disciplines use the opportunity to develop their software skills mostly through self-learning.

Once the informal information is kept aside; nothing in or related to BIM is practiced in Sri Lankan industry. As of the day of writing, no Sri Lankan building project had adopted BIM. A conscious effort to adopt BIM in recent future in the local industry was not evident. However, a considerable interest on the topic BIM was increasingly found to be discussed at professional gatherings. For this reason, Jayasena and Weddikkara (2013) labelled the Sri Lankan Construction Industry as a BIM Infant Industry - an industry that does not perform anything that is BIM, but shows an interest in adopting BIM in future.

7.2. BIM AWARENESS

Improving the BIM awareness within the industry and beyond is a significant challenge in a BIM infant industry. BIMLab Network research group centred at the University of Moratuwa was the only active organization which has taken steps to increase the BIM awareness in the country at the time of writing. They are an informal group with primary interest in research related to BIM and IPD which disseminate knowledge and information to public through their website and their email newsletter. One of their articles titled "BIM for dim" (BIMLab Network, 2014) was found to be an effective way of describing BIM to a layman to enable better understanding of what BIM in fact is.

The article starts by describing a simple building completely in words (*i.e.,* without drawings) and convinces the reader that a builder could build that building accurately by visualizing the building in his mind by interpreting this description. This phenomenon is then related to a digital description which is written in a manner that a computer can interpret accurately. In a subsequent publication they expand to explain this using a portion of a BIM (IFC) file shown in Fig. **13**.

Authors describe that the figure "shows a small portion of open standard BIM model when opened in a Notepad application. It is in plain text (symbols) which human can read and interpret. For example, #9806 describes a wall element, giving some unique information about it and refers to some other locations in the model file *viz.* #13, #9803 and #9876, where each refers to model ownership, placement of the building, and geometrical placement of the wall. A unique feature of a BIM model is that a particular information is recorded only once in the model and all other instances are reference to this information, making it easy to update information without causing any errors. A BIM model extends to millions or billions of lines making it impossible for humans to interpret them; yet they are machine (computer) readable and the computers can interpret them by using the predefined syntax and libraries" (Jayasena and De Silva, 2013, pp. Sri Lanka 4-5).

```
...
#13= IFCOWNERHISTORY(#12,#5,$,.ADDED.,$,$,$,1286451639);
...
#36= IFCDIRECTION((0.,0.,1.));
..
#9792= IFCDIRECTION((-1.,0.,0.));
#9796= IFCCARTESIANPOINT((12.,0.,0.));
#9800= IFCAXIS2PLACEMENT3D(#9796,#36,#9792);
#9803= IFCLOCALPLACEMENT(#593,#9800);
#9806= IFCWALLSTANDARDCASE('16DNNqzfP2thtfaOflvsKA',#13, 'Wand-Ext-ERDG-4',$,$,#9803,#9876,'A6C3DE63-3731-
    4F6A-94-7E-DE8A8295779F');
#9825= IFCCARTESIANPOINT((0.,0.));
#9829= IFCCARTESIANPOINT((12.,0.));
#9833= IFCPOLYLINE((#9825,#9829));
#9837= IFCSHAPEREPRESENTATION(#51,'Axis','Curve2D',(#9833));
```

Figure 13: IFC building information model. (Source: Jayasena and De Silva, 2013).

Once the computer can accurately interpret the description, by using its computational power and memory accuracy (which are now significantly higher than those of human) the computer can fully construct the building virtually in its memory. The level of details in this virtual model corresponds to the level of details in the digital description and stated reference to standards. A Building Information Model is therefore construed as a "computer interpretable digital description" of physical and functional characteristics of a facility.

Building Information Modelling is first introduced as the use cases of the Model. Use cases can be creating, modifying, sharing and using of the model. However, this understanding is short of expected understanding of what Building Information Modelling is. However, detailed discussion on each of the use case signifies the challenges and the need for changes, yielding improved awareness of BIM.

While it being different from the main stream BIM definitions, the above description is effective in introducing BIM to a BIM infant industry. The understanding is not incompatible with the mainstream BIM definitions appeared early in this book, it in fact complements by providing better understanding about BIM for BIM infants.

7.2.1. BIM Capability

It has been realized that most BIM users do not require in-depth knowledge in BIM technology. Two important aspects of BIM Capability are Technical Capability and Process Capability. Process Capability refers to the ability to work effectively in the BIM environment with the primary concern on collaborative working as required by BIM systems and BIM Execution Plans. Technical Capability refers to the ability to use relevant BIM software applications.

An interesting fact which is not often highlighted in BIM literature is, that it is not necessary to know what BIM is for a practitioner to become "BIM Ready". While knowledge on BIM and technology behind will definitely be advantageous, what is required for a practitioner is the ability to use BIM software application(s) relevant to his practice. Though ArchiCAD was used for many years, only recently it became widely known as a BIM tool. BIM software enables practitioners to work in a user-friendly computer interface, while working in the backend to perform translations (to IFC for example) and communication. Therefore, not each and every player needs to be thorough about BIM technology. One technology expert, usually identified as BIM Manager would develop and maintain the BIM system, while a BIM coordinator could coordinate with all the players to assure required performance.

7.2.2. Comparative Maturity

Three BIM maturity models were introduced earlier in this book. Sri Lankan Construction Industry is placed at "Phase 0" Bew-Richards BIM Maturity Model presented in Fig. **8**. Currently there is no CAD standards established; CPIC, BS1192, or equivalents are not practiced. Electronic exchange of CAD drawings is common. Little or no intelligence built into CAD drawings is often evidence. This context corresponds to the "Level 0" of the Collaborative BIM Decision Framework developed by Gu and London (2010). The Interactive Capability Maturity Model (I-CMM) by National Building Information Model Standard (NBIMS) Committee and buildingSMART Alliance was identified herein before to enables smoother, or more granular, progression than these 3 step-change maturity models. Therefore, it was considered worthwhile to review the Sri Lankan context using the I-CMM. Table **16** presents the perceived levels of current maturity and achievable maturity. Respective credit levels of each area of interest are presented in Figs. **14** and **15**. The most current version of I-CMM, the version 2, a downloadable spreadsheet file from National BIM Standards (NIBS) website (National Institute of Building Sciences and buildingSMART Alliance, 2012) was used for the review.

Application of I-CMM was found to be impractical without an actual BIM implementation. Therefore, the current level of maturity was identified by assuming a BIM implementation within the current practices which are known to be better than the average practices. For the review, what was considered were not the average level practices because BIM adoption is likely to occur among leaders and others will follow. Achievable maturity identified was hypothetical, and was a logical construct which is believed to be practical within 6 to 10 months as discussed below.

Table 16: I-CMM - current and achievable maturity levels

Area of Interest	Current Maturity	Achievable Maturity
Data Richness	Basic Core Data	Data Plus Expanded Information
Life-cycle Views	Planning & Design	Includes Construction/ Supply
Roles or Disciplines	Only One Role Supported	Two Roles Fully Supported
Business process	Few Business Processes Collect Information	Most Business Processes Collect Information
Delivery Method	Network Access with Basic Information Assurance	Full Web Enabled Services with Information Assurance
Timeliness/Response	Most Response Info manually re-collected	Limited Response Info Available In BIM
Change Management	No CM Capability	Aware of CM and Root Cause Analysis
Graphical Information	2D Non-Intelligent As Designed	3D - Current And Intelligent
Spatial Capability	Spatially Located	Spatially located with Metadata
Information Accuracy	No Ground Truth	Limited Comp Areas & Ground Truth
Interoperability/ IFC Support	No Interoperability	Most Info Transfers Between COTS
Credit Sum	18.2 / 100	51.2 / 100
Maturity Level	Not Certified (41.8 to Certified)	Minimum BIM (8.8 to Certified)

Source: National Building Information Model Standard (NBIMS) (2012).

7.2.3. Data Richness

In absence of practical implementation of BIM, current level of data richness, or the completeness of Building Information Models was assumed to be at the basic level. At the current level of technological standards exist in the industry, basic implementation of BIM framework for building data is practical. However, by bringing in the capable individuals who (informally) use BIM authoring applications, and by training them on necessary standards to be followed; data richness could be improved to a level that BIM would be accepted as authoritative and the primary source. This level could be envisaged because their current models are accepted indirectly as authoritative through 2D outputs they generate from 3D BIM models.

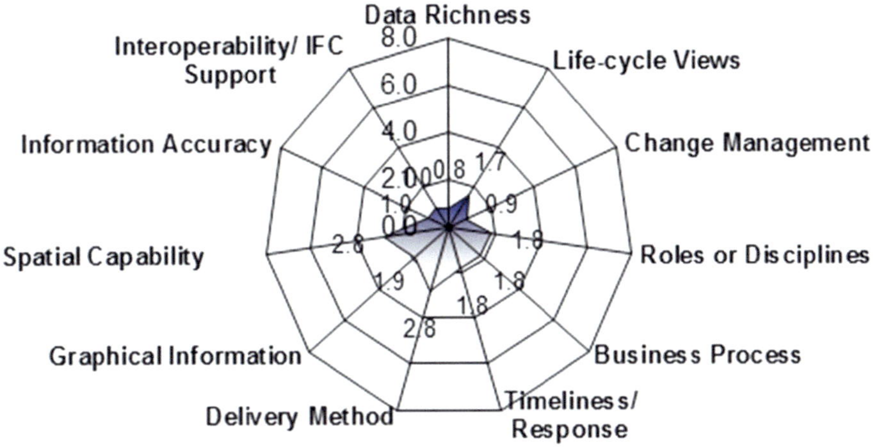

Figure 14: Current level of maturity - credit values. (Source: Developed for this book).

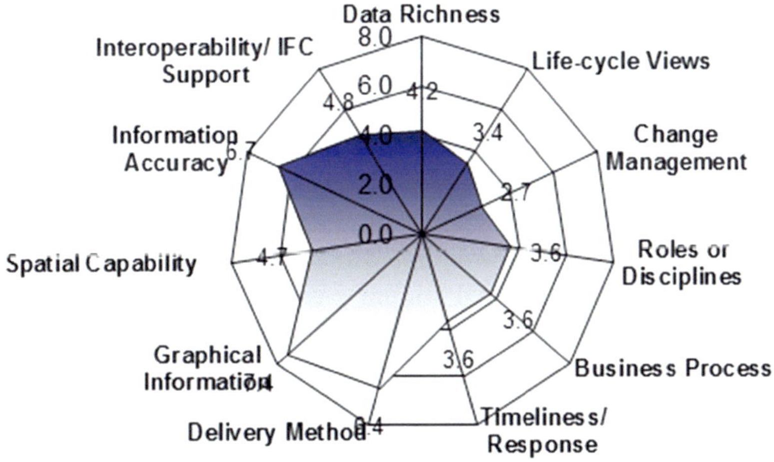

Figure 15: Achievable level of maturity - credit values. (Source: Developed for this book).

7.2.3.1. Life-Cycle Views

In current practices basic initial data is collected during planning and design; and this will be the level of maturity immediately available. Lack of collaborative working would hinder the life-cycle views being extended to next levels. However, as it will be identified herein later, collaborative working is not impossible in Sri Lankan construction industry. Facilities Management is also now a recognized discipline especially among corporate clients (de Silva, 2011). Thus with some effort, the level could be improved to include construction phase, and also the operation phase to some extent.

7.2.3.2. Roles or Disciplines

While the established BIM framework would support multiple roles, in the current context of the industry, it is unlikely that anybody other than the designer would be allowed to contribute to it. However, some of the leading builders have become very much capable in Information Management and have started to use ERP systems to manage their building information. Once the collaborative working environment is effectively established, with existing skills of builders, BIM can be extended to fully support design and construction roles.

7.2.3.3. Business Process

It is unlikely that all players in the process will become BIM ready. Some may not be willing to embrace the technology; for example, some Architects would prefer the conventional method with the belief that BIM tools would hinder the design creativity due to: parametric limitations; interoperability; and the demand for detailed information at preliminary design stages (Ahmad *et al.*, 2013). Recalling how CAD Draughtsmen bridged the gap between designers and CAD (Computer Aided Draughting) Systems; it is envisaged that a layer of BIM Modellers would bridge the gap between designers and BIM implementations in Sri Lanka (Jayasena, 2014). However, BIM modellers will be comparatively expensive, yet the projects will benefit from their higher calibre and knowledge in construction technology. In this arrangement, information will not be directly collected to the BIM through the business process. Additional cost will occur to employ BIM Modellers to add information to the BIM. However, some of the information that would be fed by Modellers (mind their higher calibre and technology knowledge) will be directly collected to the BIM. Once the professionals realize the advantages of collaborative working, up-to-date information, and accuracy; more professionals are expected work directly in BIM.

7.2.3.4. Delivery Method

At present, some designing firms have adopted network based information sharing in order to improve their work efficiency. Often password authenticated information retrieval and entry is practiced and therefore, almost all the staff at those organizations is used to work with network shared information. They will quickly adapt to similar BIM system. However, when current web-based information management practices through ERP systems by the builders is considered; with an effective collaborative system in place, the BIM can be in a web environment and multiple people would operate on it with manually controlled role based information assurance.

7.2.3.5. Timeliness/Response

Asymmetric and incomplete information has been a common problem in Sri Lankan construction industry. Designers usually communicate minimum information through documents (drawings, specification, *etc.*) and often there is incomplete information that the builder has to request for information in order to carry out the contracted works. This culture is unlikely to change immediately by implementing BIM. What can be expected from designers is a model with minimum information that is required to produce a visually complete model in BIM authoring application. However, even this minimum information will offer an improvement to present day practice. A positive change can be expected because of the transparency of the design process to the client and involvement of the builder during the design development. Parties would soon realize the value of near accurate up-to-date information for better quality decision making thus make it a part of the business processes, because it does not become a burden or significant additional work.

7.2.3.6. Change Management

A conscious methodical change management effort is not heard of in Sri Lankan construction industry. Changes are ad hoc and occur as a natural response to situations. While the responses may not be the optimal, they shall not be considered inadequate because businesses continue successfully with those changes. However, in terms of Structured Change Management, the Sri Lankan industry could be ranked only at the minimum level. At this level, it is not clear if the industry solves its business process problems by eliminating root causes or casual factors. However, identification of root causes is generally attempted in an unstructured manner. Thus the organizations would gradually embrace systematic change management if they are provided with skills, knowledge and tools.

7.2.3.7. Graphical Information

"2D Non-Intelligent as Designed Information" is the current norm in the Sri Lankan construction industry. These drawings are stored in electronic format for sharing, but there is no interaction with information. No CAD Standards are used in preparation of drawings. Immediate improvement to be expected by implementing BIM is the minimum standardization that can occur as requirement of 3D BIM authoring tools adopted. Pressure from team players, especially the Quantity Surveyors, would influence to improve the standardization. With the collaborative involvement of the builder, design and as-built information shall be made available in 3D object based graphical information and shall be kept current.

7.2.3.8. Spatial Capability

It is now quite common to spatially locate a proposed building in "Google Earth" at its concept design stage often using Trimble's SketchUp (formerly Google SketchUp) application. Thus the buildings are recognized in a world view spatially. However, the detailed designs are developed separately often using CAD tool, and therefore the spatial information is lost at this stage. Information is not shared between the BIM and GIS (geographic information system). GIS is not yet popular in construction sector. Once 3D object modelling is adopted with the BIM implementation, the benefit of spatially locating the building will be significant, and BIM authoring tools readily facilitate it. However, it is still unlikely that this information will be integrated with GIS.

7.2.3.9. Information Accuracy

At current practice building information is manually loaded to the model and no ground truthing data is made available. No electronic verification of information is employed. However, BIM applications come with advance capabilities to maintain information accuracy. Parametric object oriented modelling with relational data structures enable a change in a parameter of one object to be reflected in all other relevant objects and spaces in the model. Therefore, employing of BIM authoring tools for design and model buildings itself will improve the Information Accuracy. Improved accuracy will increase the acceptance of BIM systems. However, there may be certain limitations in acceptance; for example, Quantity Surveyors may be reluctant to rely on automated quantity take-off (Smith, 2014, Wijayakumar and Jayasena, 2013, Olatunji *et al.*, 2010).

7.2.3.10. Interoperability and IFC Support

The ultimate goal of BIM is the interoperability of building information. It is common for BIM maturity models full interoperability at the highest level of maturity. Thus, the use of open Standard IFC is often attributed to this level. As it will be discussed in following section, adopting IFC at a lower maturity level is not impractical. However, proprietary BIM systems are the popular solutions for early BIM adoption. Products from the same vendor offer better interoperability, however not all the software tools used in the projects may be supported.

7.2.4. Section Summary

The review suggests that the BIM maturity of the Sri Lankan construction industry is very low. With a low I-CMM credit sum of 18.2 and 41.8 more credits to earn, it is identified as "not certified". Thus, at current level of BIM maturity, forming a team to implement a project using BIM is not practical. The forecast of short-term achievement of BIM maturity had a credit score of 51.2 with 8.8 credits short to reach "certified" level. This level of maturity is identified minimum BIM. The findings suggest that an industry wide structured endeavour is necessary to improve its BIM maturity to an acceptable level.

7.3. BIM INTEGRATION INTO CURRENT PROCUREMENT SYSTEMS

The procurement system is in fact another perspective of the project strategy that is developed to achieve the project objectives. It is usually a Temporary Multi Organization (TMO), an organized, purposeful structure regarded as a whole and consisting of interrelated and interdependent elements. The elements are the team members, and are usually other organizations represented by authorized persons. This structure entails a complex set of relationships between parties with different professional backgrounds. The system complexity is further increased by complexity of the goal it tries to achieve. Buildings are unique and complex one-off products (Wild, 2001). Procurement systems have been designed to maximize the system homeostasis to ensure expected (or near expected) performance. Selection of team members and appropriate contracts are key aspects of establishing effective procurement systems.

7.3.1. BIM and Standard Forms of Contract (Gap and Filling the Gap)

It is clear that BIM is a developing scientific and procedural change within the construction industry (Eastman *et al.*, 2011, Succar, 2009). With this change, how effective would the conventional forms of contract be, is a question being asked. Several standard forms are developed to facilitate project implementation with BIM. Among notable forms are AIA C191 (Standard Form of Agreement

Between Owner and Contractor for Integrated Project Delivery), AIA A295 (General Conditions of the Contract for Integrated Project Delivery), and ConsensusDOCS 300 (Tri-Party Agreement for Integrated Project Delivery) with 301 BIM Addendum; all with United States origin. Chartered Institute of Building, United Kingdom, published Complex Projects Contract (CPC) in 2013 to facilitate project implementation with BIM (Pickavance, 2013).

The conventional standard forms of contracts used in Sri Lanka are not readily suitable for implementing projects using BIM. The incompatibility arise because BIM makes a paradigm shift in the way building information is created, recorded and communicated. The contract adopted should support the new information approach. However, a total deviation from familiar standard forms would be a significant change that the industry cannot effectively respond. Thus a solution like UK BIM Protocol which directly addresses the need, and the need only, becomes the most practical solution.

A general principle of BIM Protocol is that "the Protocol is a contractual document which takes precedence over existing agreements. The primary objective of the Protocol is to enable the production of Building Information Models at defined stages of a project. The Protocol incorporates provisions which support the production of deliverables for 'data drops' at defined project stages. The Protocol also provides for the appointment of an Information Manager" (BIM Task Group, 2013, p. iv). Thus the protocol is used on top of an existing form of contract. Thus it will make the minimum necessary change to the current process. "A further objective of the Protocol is that its use will support the adoption of effective collaborative working practices in Project Teams" (BIM Task Group, 2013, p. iv). The guidance notes further states that "the Protocol is intended to be expressly incorporated into all direct contracts between the Employer and the Project Team Members" (BIM Task Group, 2013, p. v) creating an effective multiparty agreement for the BIM.

In Priority of Documents the Protocol states "in the event of any conflict or inconsistency between a Model prepared and delivered in accordance with this Protocol and any document or information extracted from such Model, except where the Information Requirements states otherwise, the Model shall prevail" (BIM Task Group, 2013, p. 2). This clause makes the paradigm shift in building information to BIM from conventional methods with minimum or no change to the roles of team members. Therefore, adoption of UK BIM Protocol on top of conventional standard form of contracts is the most practical solution identified for the Sri Lankan construction industry.

7.3.2. Collaborative Working

A contractual arrangement is not sufficient to promote collaborative working. The concept of "Relational Contracting" is often used to improve collaborative working. The concept covers a range of delivery methods emphasize and focus upon the relationship among team members. They are informal agreements and codes of conduct among the members sustained by the value of relationships (Chan *et al.*, 2004).

Relational Contracting is not reported from Sri Lankan industry. While it has been promoted by many academic writing for more than two decades, no single attempt has been made. However, this does not suggest that such collaborative approaches are impossible in the Sri Lankan construction industry. If one studies only the relationships in a project implemented with Conventional Method would strongly believe that Relational Contracting is not possible. Nonetheless, Gunathilake and Jayasena (2008) have shown that the attitudes and culture of team members changed favourably to collaborative working once the project environment is changed to a procurement system that supports and promotes collaboration. Interestingly, there were few projects executed without a formal contract. Thus the evidence suggests, if properly managed, Relational Contracting is possible in the Sri Lankan industry.

7.3.3. Legal Context

With the change in the working process and methods, there will be legal uncertainties primarily in terms of design liability and intellectual property rights of the design. The legal position and relationships are likely to have major deviation from those of conventional practices is only at the highest maturity stage *i.e.,* when moving to iBIM (integrated BIM). At this stage, there will be a single model which is accessed and updated by all team members. However, deviations can occur at lower maturity levels also because of the changes in process. For example, liability of fulfilling time and quality targets of 'data drops' becomes significantly high because not like human brain, computers are less tolerant to missing and erroneous data. Since BIM relies on data interpreted by computers, the overall impact of poor performance of a member will be significantly high. The gravity of the legal responsibility may not be readily understood by the members. Therefore, a need to educate the team members on the change of legal context exists.

Intellectual property (IP) in Sri Lanka is governed by Intellectual Property Act No 36 of 2003. "IP refers to creations of the mind, such as inventions; literary and artistic works; designs; and symbols, names and images used in commerce. IP is protected in law by, for example, patents, copyright and trademarks, which enable people to earn recognition or financial benefit from what they invent or create". (World Intellectual Property Organization, n.d.). Therefore, not only the financial benefit, but recognition is also protected by IP. When the model is centrally developed with contribution of many parties, what constitutes the IP, and who owns that becomes a question. These issues need to be addressed. The UK BIM protocol Clause 6 "sets out the IP provisions required to enable the Models to be used as intended and to protect the rights of the Project Team Members against infringement" (BIM Task Group, 2013, p.v).

7.3.4. Section Summary

A procurement system is an organized, purposeful structure regarded as a whole and consisting of interrelated and interdependent elements created to get a facility delivered successfully. Few procurement systems have become standard systems, or methods. Conventional Method (also known as Traditional Method) is the dominating procurement method in Sri Lanka. Design and Build, Package Deal and Turnkey methods are also found but not common. Relational approaches are not heard of. The context signifies the reluctance of the industry to change.

A BIM implementation, however, will require a change from traditional methods. While there are alternative contracts available, what is preferred for Sri Lankan context is the UK BIM Protocol on top of conventional form of contract in use. This will require minimum change to existing methods while creating the necessary contractual context to collaborative working. Additional endeavour to create Relational Contracting can be used to improve the context.

7.4. BIM SOFTWARE ENVIRONMENT

A wide variety of software has been developed for BIM since its early concepts were adopted in building industry more than two decades ago. Arbitrary selection of software can give rise to issues in terms of integration, consistency, accuracy and affordability. To minimize potential issues, an assessment of varying software environments for their suitability for Sri Lankan context is necessary.

Among the software applications widely used in the industry are Autodesk Revit, Graphisoft ArchiCAD, Bentley Architecture, Innovaya Visual BIM, Tekla

Structures, Exactal CostX, Vico Estimator, Synchro, Autodesk Navisworks, and Solibri IFC Model Checker (Jayasena and Weddikkara, 2014). Each application is developed to be used by one of the key disciplines in the industry; while some may be used by more than one discipline. Following is a list of factors that would affect the choice of software application. (Khemlani, 2012, Arayici *et al.*, 2011, Luthra, 2010, Gunasekara and Jayasena, 2013, Won and Lee, 2010, Eastman *et al.*, 2011).

1. Modelling and viewing capabilities of Application
2. Accuracy of data in models
3. Sharing capability of models
4. Information documenting capabilities
5. Spread of usage or popularity of Application
6. Affordability

The above list can be use as a guide to review the selection of software applications. However, the selection cannot be done independently because incompatibilities with the applications used by other team members can occur. Therefore, it becomes wise to consider overall BIM system, inclusive of software applications, model storage and communication which is identified as the BIM Software Environment (BIMSE).

7.4.1. Classification of BIMSE

Since there can be a number of different BIMSEs, classification of them will help to offer better understanding and guidance for selection. For the purpose of early decision making, a classification based on data model as (1) Homogeneous Software Environment (HSE), and (2) Plural Software Environment (PSE) is most suitable. A comparison between two BIMSEs is presented in Table **17**.

Table 17: Homogeneous and plural software environments - comparison (Source: Gunasekara and Jayasena, 2013)

Criterion	Homogeneous Software Environment	Plural Software Environment
Data sharing platform	Central data repository	Shared or central data repository
BIM Software	Proprietary applications/suites - Forced to	Bespoke middleware created for IFC

Criterion	Homogeneous Software Environment	Plural Software Environment
	select the same software by all collaborating parties. Cost of software is high and license sharing capability varies.	based applications - Collaborating parties are free to choose own software tools to achieve higher performance in their AEC tasks
Data exchange	Primarily based on file formats that depend on the selected software	Primarily based on IFC data model
Data amalgamation/ Data fusion	Performed *via* live synchronization of data. Data duplication is possible	Performed *via* model fusion with fusion algorithms Data duplication is possible
BIM software modelling expertise	All team members require equal expertise	All team members do not require equal expertise when using reference models and IFC
Selection of project partners	Focus on the competency of a specific BIM software tool may overlook the actual competency in performing engineering tasks.	The use of reference model concept with IFC reduces the needed competency in BIM, there by maintaining the requirement of actual competency in performing engineering tasks.

7.4.2. Homogeneous Software Environment

In an HSE, one application or a suite of applications from one software vendor is identified to be the central software application. Applications so selected usually are also capable of performing as the model server application, or are supplied with a separate server module (Autodesk, n.d.-a, Graphisoft, n.d.-a, Bentley, 2014) which can play the role of Central Data Repository. Understanding HSE using Jernigan's (2008) "BIG BIM, little BIM" would improve the comprehension. Little BIM is the BIM within the organizational boundary of a team member (*i.e.,* the organization) of TMO. BIG BIM is what is beyond this boundary. In the HSE, the BIG BIM represents the server which holds the Central Building Information Model in the selected HSE native format. All team members' little BIMs should be compatible with the selected HSE native format. Thus HSE is synonymous to pBIM (proprietary BIM) identified earlier in this book.

Fig. **16** is a graphical illustration of HSE. As depicted in it, majority of little BIMs will work with common native format. However, some little BIMs, such as Quantity Surveying (QS), would use exchange formats offered by pBIM systems.

7.4.3. Plural Software Environment

PSE does not limit the software application selection based on a native format. Instead, it relies on open standard IFC data exchange and offer freedom to select any application with IFC compliance. In the PSE, little BIM's do not work in IFC format. Instead they work in the native format of the software they use. The native format of each little BIM needs not to be similar. The software in little BIM translates the model in native format to a copy in IFC when it needs to be communicated to another team member. Software in the receiving little BIM will translate the IFC into its native format and continue to develop the model. Once done, a copy of it in IFC format will be sent back to the other team member who can update his own native format model with new data in IFC file. Therefore, in a PSE, a central model server is not required. The context is presented in Fig. **17**. As illustrated in it, there is no single model that can be referred to as the BIG BIM. BIG BIM is the 'cloud' among the little BIMs. What it constitutes is undefined; it can be a Local Area Network (LAN), Wide Area Network (WAN), or Cloud Storage.

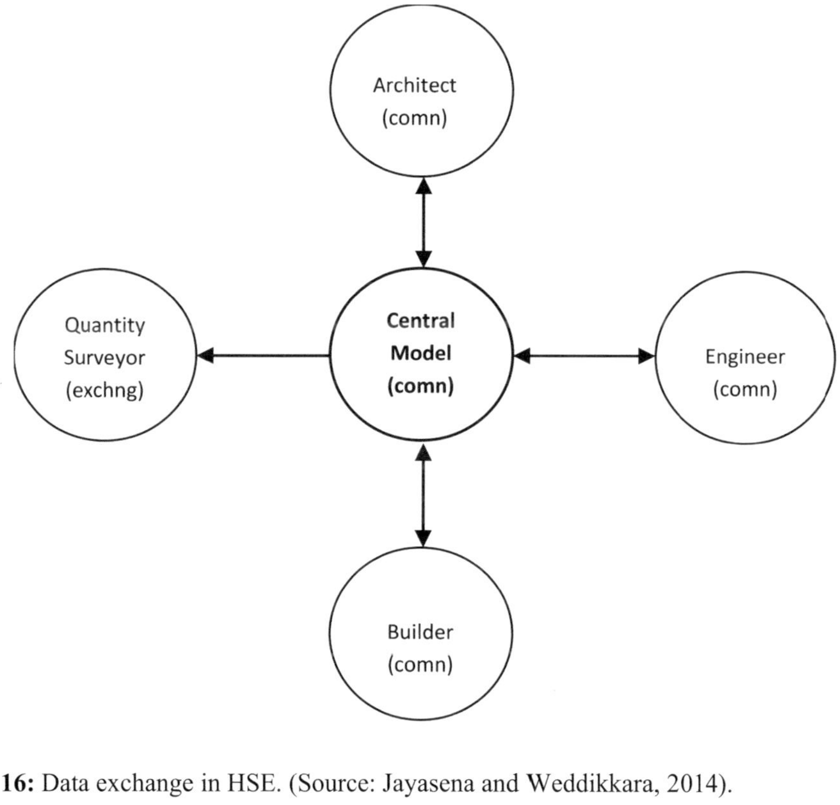

Figure 16: Data exchange in HSE. (Source: Jayasena and Weddikkara, 2014).

van Berlo *et al.,* (2012) reported and experimental study of BIM implementation using Central BIM Server and IFC information exchange. Thus, it becomes an improved version of PSE. open source BIM server (refer BiMserver, n.d.) was used as the Central Data Repository. They share that "it is true that IFC will not contain the full dataset of the original, native data-base from the source software, but often the receiving user does not require all details. Occasionally some data became unreliable (objects are misplaced or gone) during import, but the overall quality of contemporary implementations was considered satisfying. The imported data is used as a reference during engineering. For example the MEP engineer uses some data from the structural engineer to design the location of the piping. The piping is not added to the original dataset of the structural engineer, but exported to IFC as a new dataset that is send to the central data repository" (van Berlo *et al.*, 2012).

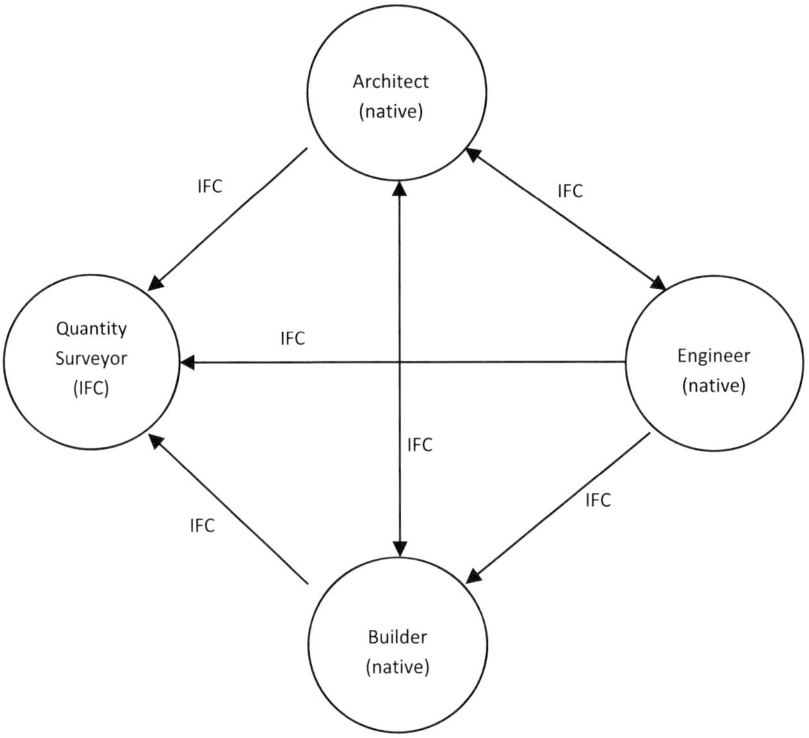

Figure 17: Data exchange in PSE. (Source: Jayasena and Weddikkara, 2014).

7.4.3. Analysis of BIMSE for Sri Lankan Context

Suitable BIMSE for Sri Lankan context have been reviewed by Jayasena and Weddikkara (2014). In absence of actual BIM implementations in the country, their review was primarily based on literature synthesis. They conclude that "that Plural Software Environment likely to be the preferred solution for Sri Lankan context. It will offer the flexibility for participants to select the software tools for their requirements, preference and affordability. Open source BIM Server, which is freely available, is recommended to be used as the central data repository. Implementation will obviously require the support of Information Technology Experts" (Jayasena and Weddikkara, 2014, p. 105).

They make this conclusion by analysing several aspects of BIM with the Sri Lankan context. They use the gaining popularity of open BIM standard (NBS, 2014) and continued development of it (Liebich *et al.*, 2013) to further strengthen it. They report that "BIM practices are not totally alien to the Sri Lankan construction industry. Though there are no clear reports, it is known that BIM capable software such as Revit, Archicad and CostX being used by some in the industry. However, these uses are not in BIM based project environment, but used as in-house tools to generate non-BIM information such as 2D drawings and 3D graphics" (Jayasena and Weddikkara, 2014, p. 104).

They agree that "BIM maturing and infant industries are likely to prefer Homogeneous Software Environment offered by software vendors as proprietary solutions due to accompanied customer support including structured learning resources" (Jayasena and Weddikkara, 2014, p. 104). Yet express their concern on overall effectiveness of such selections primarily in terms of cost effectiveness and acceptance. They argue that "Multiparty collaborating using a central model may not be readily acceptable for the industry participants. Members of TMO will prefer to use private repositories (or models) to keep their unshared data within the little BIM, and as the design (or model) develops exchange the permissible data with the BIG BIM (Nour, 2009). Ideally, the software tools must be selected based on the tasks performed by the member, but not on the ability to share data with others. In an industry of a developing economy, affordability will be a significant factor in selection of software applications. Plural Software Environment will allow the flexibility, and the increasing support for IFC with the development of free and/or open source software will make the technology affordable for participants with limited affordability (budget constraints)" (Jayasena and Weddikkara, 2014, pp. 104-105).

One factor authors had not considered was the potential role of Information and Communication Technology Agency (ICTA) of Sri Lanka. "Sri Lanka has shot up to 74th position in the United Nations E-Government Survey of 2014, after climbing 41 places since 2012" (Information and Communication Technology Agency, 2014). This is evidence of significant effort by ICTA to develop the national ICT infrastructure and promote e-Governance. ICTA also manages the interoperability initiative of the Sri Lanka government. Current initiative includes (1) Land Domain, (2) Vehicle Domain, (3) Personal Information Domain, and (4) Foreign Funded Project Coordination Domain. Many other initiatives of ICTA show their keen interest in mobilizing Information and Communication technology for the benefit and development of the country. There was no evidence that BIM could be in ICTA's agenda. However, if convinced, due to the reported advantages of BIM, ICTA is likely to take initiatives supporting its adoption.

BIM initiation by an organization like ICTA would significantly change the BIM adoption strategies. Since it possesses infrastructure and capacities, it can offer cloud-based BIM systems on rental, probably at subsidized rates. This would change the industry concern of affordability, and proprietary BIM systems may become the preferred choice. Knowing the current deficiencies in operation of public buildings, getting new public building procured using BIM would become a preferred choice. Thus it will help faster uptake of BIM in the country.

7.4.4. Section Summary

A wide variety of BIM software is made available by software vendors. Selection of software applications is primarily affected by the features of the software and the affordability. However, due to need of efficient information sharing and interoperability, the BIM Software Environment (BIMSE) would first affect the selection of software; or the selection of team members will depend on what software they use. Therefore, the decision of BIMSE becomes critical.

Two distinct types of BIMSEs were identified. They are (1) Homogeneous Software Environment (HSE), and (2) Plural Software Environment (PSE). As a generic conclusion, PSE had previously being identified as the suitable BIMSE for Sri Lankan industry, and found to be acceptable with one exception. Active involvement of Information and Communication Technology Agency (ICTA) of Sri Lanka would significantly change the context so that preferred BIMSE would shift to HSE. However, BIM is not in current ICTA agenda. An effort to convince ICTA of the benefit the country could receive by implementing BIM would become a significant leap in BIM adoption in the country.

CHAPTER 8

Reflection and Future BIM Research

Abstract: The technology underpinning BIM including hardware, BIM software, facilitating hardware such as IT infrastructure (firm, national and international, fixed line and mobile), software such as collaboration platforms, document management systems and emerging cloud based technologies, continues to rapidly evolve and converge. Capability is continuing to evolve with BIM data now being integrated with site surveying and engineering equipment, such as total stations, helping BIM move to site. RFID tagging and bar-coding are amongst the many initiatives extending capability throughout the supply chain. As such, the awareness and adoption of BIM can be regarded in a longitudinal perspective. The human change resistance perspective relating to acceptance of new technology and legal implications of BIM implementation are recommended for further investigations. Besides, currently there is no reliable research regarding the adoption of BIM amongst different countries and relative to each other. It is recommended that comparison research is carried out in this area. This book provides a useful insight on BIM in Malaysia and Sri Lanka and also serves as a practical reference for the developing countries.

The book describes BIM evolution in two different countries from developing nations. BIM is getting more popular for both countries; however the actual use of BIM can be seen in Malaysia at the moment.

The technology underpinning BIM including hardware, BIM software, facilitating hardware such as IT infrastructure (firm, national and international, fixed line and mobile), softwares such as collaboration platforms, document management systems and emerging cloud based technologies, continues to rapidly evolve and converge. This evolution is making Level 2 BIM more readily accessible in Malaysia soon.

The software is becoming more intelligent and continuous development of the user interface should make it easier to master and more accessible, this is particularly true as collaboration platforms and document management systems are beginning to become possible to integrate into one comprehensive and holistic solution, although much development work needs to be done. Capability is continuing to evolve with BIM data now being integrated with site surveying and engineering equipment, such as total stations, helping BIM move to site. RFID tagging and bar-coding are amongst the many initiatives extending capability throughout the supply chain. As such, the awareness and adoption of BIM can be

John Rogers, Heap-Yih Chong, Christopher Preece, Chai Chai Lim and Himal Suranga Jayasena

regarded in a longitudinal perspective. This book is a cross-sectional study and provides a point of departure for longitudinal research into the same topic area, with particular focus upon any emergent government policy initiatives that support BIM adoption.

BIM is inherently collaborative and requires the participation of many stakeholder groups. Indeed it is likely that the greater involvement with the value chain the greater the benefits that can be achieved through BIM. As such, continuing research is necessary among key stakeholder groups in Malaysia; design team, especially Architects as lead designer and base model producer, quantity surveyors, planners, estate managers, clients, contractors, government departments.

The adoption of BIM is a strategic decision for a firm. Firms need feedback on their strategic decisions to confirm or amend their earlier decisions. This can extend across a range of indices, although typically for a firm this will include return-on-investment analysis and market positioning. As yet, there are no reliable metrics known for establishing the cost-benefit of adopting BIM. The construction of reliable metrics would help make assessment of BIM adoption more objective. Development of these metrics would inform strategic decision making. This represents opportunity for further research.

BIM impacts across a wide range of an adopting firm's business activities, including external business relationships. External business relationships are subject to Law. Notable concerns exist within the Malaysian AECO sectors pertaining to the legal implications of BIM. It is recommended that further, specialist research is carried out in this area.

The human change resistance perspective relating to acceptance of new technology is an important component of technology adoption. It is recommended that further, specialist research is carried out in this area. Research pertaining to related human resource issues, including staff retention, would also prove valuable.

BIM is a global phenomenon. Much of the development of the core processes and core software has taken place at a global level, albeit with some minor modification for different countries or regions. Currently there is no reliable research regarding the adoption of BIM amongst different countries and relative to each other. It is recommended that comparison research is carried out in this area.

Although BIM is beginning to demonstrate its early potential as evidenced by the successful completion of projects, it should not be concluded that BIM is suitable for all parties, projects or participants at all times. The level of maturity that is appropriate for a broad spectrum of parties, projects or participants would benefit from further research.

In conclusion, this book provides a useful insight on BIM in Malaysia and Sri Lanka and also serves as a practical reference for the developing countries. Readers will appreciate the fundamental approach of the discussion on the growing BIM body of knowledge, philosophies and trends.

REFERENCES

REFERENCES

Ab Halim, M. S., Jaafar, M. & Osman, O. (2011). Assessment of the Financial Health of Malaysian Construction Firms using Financial Ratio Analysis. *Internation Journal of Academic Research,* 3, 200-207.

Abdelhamid, T. S. (2004). The self-destruction and renewal of lean construction theory: a prediction from Boyd's Theory'.

Abdul-Rahman, H., Berawi, M. A., Berawi, A. R., Mohamed, O., Othman, M. & Yahya, I. A. (2006). Delay Mitigation in the Malaysian Construction Industry. *Journal of Construction Engineering and Management,* 132, 125-125.

Abdullah, M. & Egbu, C. (2010). The role of knowledge management in improving the adoption and implementation practices of industrialised building system (IBS) in Malaysia. School of Built Environment, The University of Salford.

Abdullah, S., Razak, A., Bakar, A., Hassan, A. & Sarrazin, I. (2009). Towards Producing Best Practice in the Malaysian Construction Industry: The Barriers in Implementing the Lean Construction Approach. *Building*, 1-15.

Abidin, N. Z. (2009). Sustainable Construction in Malaysia – Developers ' Awareness. *Engineering and Technology*, 807-814.

Abidin, N. Z. (2010a). Investigating the awareness and application of sustainable construction concept by Malaysian developers. *Habitat International,* 34, 421-426.

Abidin, N. Z. (2010b). Sustainable Construction Practices in Malaysia. *In:* DINCER, I., HEPBASLI, A., Midilli, A. & Karakoc, T. H. (eds.). Springer US.

Abidin, N. Z. & Jaapar, A. (2009). Sustainable Concept Awareness In Malaysia Construction Practices. *jmu.ac.uk*, 1-9.

ACEM (2011). ACEM Directory 2010. *Association of Consulting Engineers Malaysia Secretariat,* Kuala Lumpur.

ACEM. (2012a). *Events* [Online].

ACEM. (2012b). *Registration for ordinary membership - Extracts from Articles of Association* [Online]. Available: http://www.acem.com.my/index.php?option=com_docman&task=doc_view&gid=8&Itemid=44 2012].

AEC-(UK) (2012a). BIM Protocol *v2.0.*

AEC-(UK) (2012b). BIM Protocol for Autodesk Revit. *Additional Protocol.*

AEC-(UK) (2012c). BIM Protocol Project BIM Execution Plan. *A pro-forma and guidance document to developing a Project BIM Execution Plan.*

Ahmad, A., Demian, P. & Price, A. (2013). Creativity with Building Information Modelling Tools. *International Journal of 3-D Information Modeling (IJ3DIM),* 2, 1-10.

Alinaitwe, H. M. (2009). Prioritising Lean Construction Barriers in Uganda's Construction Industry. *Journal of Construction in Developing Countries,* 14, 15-30.

American-Institute-Of-Architects. (2012a). *2012 BIM Awards* [Online]. Available: http://network.aia.org/TechnologyinArchitecturalPractice/Home/BuildingInformationModelingAwardsProgram/2012Recipients [Accessed 7.1.13.]

American-Institute-Of-Architects (2012b). AIA Document E203™–2012, Building Information Modeling and Digital Data Exhibit.

American-Institute-Of-Architects (2012c). AIA Document G201™–2012, Project Digital Data Protocol Form.

American-Institute-Of-Architects (2012d). AIA Document G202™–2012, Project Building Information Modeling Protocol Form.

American-Institute-Of-Architects (2012e). Survey of Building Information Modeling (BIM) Standards and Guidelines.

Amit, R. & Schoemaker, P. J. H. (1993). Strategic assets and organizational rent. *Strategic Management Journal,* 14, 33-46.

Arasaratnam, S. (2014). Sri Lanka: History [Online]. Encyclopædia Britannica, Inc. Available: http://www.britannica.com/EBchecked/topic/561906/Sri-Lanka/214602/History [Accessed 7 Sep 2014].

John Rogers, Heap-Yih Chong, Christopher Preece, Chai Chai Lim and Himal Suranga Jayasena

Arayici, Y., Coates, P., Koskela, L., Kagioglou, M., Usher, C. & O'reilly, K. (2010). Technology adoption in the BIM implementation for lean architectural practice. *Automation in Construction.*

Arayici, Y., Coates, P., Koskela, L., Kagioglou, M., Usher, C. & O'reilly, K. (2011). BIM adoption and implementation for architectural practices. *Structural Survey,* 29, 7-25.

Arayici, Y., Egbu, C. & Coates, P. (2012). Building information modelling (BIM) implementation and remote construction projects: issues, challenges, and critiques. *Journal of Information Technology in Construction,* 17, 75-92.

Austen, R. S., Seymour, R., Brown, K., Furneaux, C., Mccabe, A. & Bucknall, R. L. (2007). Multi-Outcome Construction Policies: Literature Review on Stakeholder Theory.

Autodesk. n.d.-a. *About Revit Server* [Online]. Available: http://knowledge.autodesk.com/support/revit-products/learn-explore/caas/CloudHelp/cloudhelp/2015/ENU/Revit-Server/files/GUID-4DE9A327-583E-40DA-81EB-1150D60E6887-htm.html [Accessed 20 April 2014].

Autodesk. n.d.-b. *Autodesk software for students and educators* [Online]. Autodesk. Available: http://www.autodesk.com/education/free-software/all [Accessed 20 Aug 2014].

Autodesk (2011). Realizing the Benefits of BIM. USA.

Autodesk. (2012a). *COBie Toolkit for Revit* [Online]. Available: http://usa.autodesk.com/adsk/servlet/oc/offer/form?siteID=123112&id=18070390 [Accessed 12.12.12.

Autodesk. (2012b). *Forum - Discussion Groups* [Online]. Available: http://forums.autodesk.com/t5/Autodesk-Revit-Architecture/Revit-Backwards-Compatibility/td-p/.UOps_2-E1HQ [Accessed 5.11.12.

Autodesk. (2013). *Autodesk Building Design Suite* [Online]. Available: http://usa.autodesk.com/autodesk-building-design-suite/compare/.

Baba, H. D. (2010). *Building information modeling in local construction industry.* Masters, Universiti Teknologi Malaysia.

Baccarini, D. (1996). The concept of project complexity--a review. *International Journal of Project Management,* 14, 201-204.

Ballard, G. (2008). The lean project delivery system: An update. *Lean Construction Journal,* 4, 1–19.

Ballard, G. & Howell, G. 1997. Implementing lean construction: stabilizing work flow. *Lean Construction*, 101-110.

Ballard, G. & Howell, G. (1998). What kind of production is construction.

Ballard, H. G. (2000). *The last planner system of production control.* Citeseer.

Barney, J. (1991). Firm resources and sustained competitive advantage. *Journal of management,* 17, 99-120.

Barney, J. B. (1986). Strategic factor markets: expectations, luck, and business strategy. *Management science*, 1231-1241.

Barrett, P. (2005). Revaluing construction: a global CIB agenda. *International Council for Research and Innovation in Building and Construction (CIB), Rotterdam, the Netherlands.*

Barrett, P. (2007). Revaluing Construction: a holistic model. *Building Research & Information,* 35, 268-286.

Bashir, A. M., Suresh, S. & Proverbs, D. G. (2010). Barriers Towards The Sustainable Implementation Of Lean Construction In The United Kingdom Construction Organisations. *ARCOM Doctoral.*

Baum, J. A. C., Dobrev, S. D. & Van Witteloostuijn, A. (2006). *Ecology and strategy*, Elsevier.

Baye, M. R. (2006). *Managerial economics and business strategy*, Irwin/McGraw-Hill.

Becerik-Gerber, B. & Rice, S. (2010). The perceived value of building information modeling in the US building industry. *ITcon,* 15, 185-201.

BEM. (2012a). *About Us* [Online]. Available: http://www.bem.org.my/v3/aboutus.html.

BEM. (2012b). *Registration Statistics* [Online]. Available: http://www.bem.org.my/v3/index.html.

BEM. (2012c). *Regulations* [Online]. Available: http://www.bem.org.my/v3/pdf/regulations_2003/s24a.pdf.

Bentley. (2014). *ProjectWise Information Management and Collaboration* [Online]. Available: http://www.bentley.com/en-US/Products/projectwise+project+team+collaboration/ [Accessed 20 April 2014].

Bentley Systems. n.d. *STUDENTserver* [Online]. Available: http://apps.bentley.com/studentserver/home/index [Accessed 20 Aug 2014].

Berard, O. & Karlshoej, J. (2012). Information delivery manuals to integrate building product information into design. ITcon.

Bernstein, P. G. & Pittman, J. H., (2005), 'Barriers to the adoption of Building Information Modeling In The Building Industry. Autodesk Building Solutions', Autodesk Inc., San Rafael, CA.

Bertelsen, N. H., Frandsen, A. K., Haugbølle, K., Huovila, P., Hansson, B. & Karud, O. J. (2010). CREDIT Performance Indicator Framework: A proposal based on studies of building cases, regulations, standards and research in seven Nordic and Baltic countries. CREDIT Report 3.

Bertelsen, S. (2004). Lean Construction: Where are we and how to proceed? *Lean Construction Journal,* 1, 46-69.

Bertelsen, S. & Koskela, L. Construction Beyond Lean: a new understanding of construction management. (2004).

Best, R. I. (2010). *Modern Construction Economics: Theory and Application*, Taylor & Francis.

Bew, M. & Underwood, J. (2009). Delivering BIM to the UK Market. *Handbook of Research on Building Information Modeling and Construction Informatics: Concepts and Technologies,* IGI-Global, Hershey, Pennsylvania, USA, pp.30-64.

BIM-Industry-Working-Group. (2011). BIM Strategy Paper for the Government Construction Client Group. London.

BIMLAB Network. (2014). *BIM for dim* [Online]. Available: http://www.bimlab.net/the-world-of-bim/more-articles/40-bim-for-dim.html [Accessed 12 Aug 2014].

BIMSERVER. n.d. *Features* [Online]. Available: http://bimserver.org/about/base-features [Accessed 23 April 2014].

BIM-SERVER. (2012). Available: http://bimserver.org/.

BIM-Task-Group. (2013). Available: http://www.bimtaskgroup.org/.

Board-Of-Architects-Malaysia. (2012). *The functions of the Board* [Online]. Available: http://www.lam.gov.my/aboutus.html.

Board-Of-Engineers-Malaysia (2012).

Bongiorni, M. J. & Cohort, I. (2011). Integrated Project Delivery.

Bower, J. L. & Christensen, C. M. (1995). Disruptive Technologies: Catching the Wave. *Harvard Business Review,* January–February (1995).

Brewer, G., Gajendran, T., Beard, C., (2011). Influences on the adoption of BPM/BIM: an Australian perspective. *CRC Press.*

Brewer, G., Gajendran, T., Runeson G Building information modeling and construction project team integration; a case of theory and practice. *In:* C. N. PREECE, F., S.M., PADFIELD, R., Papargyropoulou, E., Abdullah, M.M.B., Rogers, J., ed. International Construction Business & Management Symposium, (2011).

British-Standards-Institution (2008a). BS 1192: 2007 - Collaborative Production of Architectural, Engineering and Construction Information.

British-Standards-Institution (2008b). PAS 55 - Optimal management of physical assets.

Bryman, A. (2008). *Social Research Methods,* New York, Oxford University Press Inc.

Buildingsmart-Australasia (2012). Strategy: A strategy for the focussed adoption of building information modelling and related digital technologies and processes for the Australian built environment sector. *National Building Information Modelling Initiative. Vol.1.*

Buildingsmart. (2012a). Available: http://buildingsmart.org/.

Buildingsmart (2012b). Executive Committee Agreed Description.

Buildingsmart. (2012c). *IFC Wiki* [Online]. Available: http://www.ifcwiki.org/index.php/Main_Page [Accessed 10.12.12.

Buildingsmart (2012d). An Integrated Process for Delivering IFC Based Data Exchange

Carmona, M., Carmona, S. & Clarke, W. (2001). A bibliography of design value. London: Commission for Architecture and the Built Environment (CABE).(available at: http://www. cabe. org. uk).

Carmona, M., *et al.*: Commission For Architecture And The Built Environment (CABE) 2001. the value of urban design.

Central Bank Of Sri Lanka (2014). Economic and Social Statistics of Sri Lanka 2014, Colombo, Sri Lanka, Central Bank of Sri Lanka.

Chan, A. P., Chan, D. W., Chiang, Y., Tang, B., Chan, E. H. & Ho, K. S. (2004). Exploring critical success factors for partnering in construction projects. *Journal of Construction Engineering and Management,* 130, 188-198.

Chan, A. P. C., Fan, L. C. N. & Yu, A. T. W. (1999). Construction process reengineering: a case study. *Logistics Information Management,* 12**,** 467-476.

Chan, T. K. (2009). Measuring performance of the Malaysian construction industry. *Construction Management & Economics,* 27**,** 1231-1244.

Che Mat, M. M. (2010). Value Management - The Way Forward. *In:* CIDB (ed.). Kuala Lumpur.

Chia, F. C., Skitmore, M., Runeson, G. & Bridge, A. An assessment of construction labour productivity in Malaysia. (2010). CIB.

Cholakis, P. (2012). *BIM, Big Data, and Efficient Life-cycle Management of the Built Environment* [Online]. Available: http://buildinginformationmanagement.wordpress.com/2012/08/06/bim-big-data-and-efficient-life-cycle-management-of-the-built-environment/ [Accessed 6.9.12.

Choo, C. W. & Bontis, N. (2002). *The strategic management of intellectual capital and organizational knowledge*, Oxford University Press, USA.

Chong, H.Y., Wong, J. S., & Wang, X. (2014). An explanatory case study on cloud computing applications in the built environment. *Automation in Construction*, *44*, 152-162.

Chong, H.Y, Balamuralithara, B. And Chong. S.C. (2011). Construction contract administration in Malaysia using DFD: a conceptual model. *Industrial Management and Data Systems* 111 (9): 1449-1464.

Chong, H.Y, & Zin. R.M. (2011). Transforming of Construction Contract Administration: Data Warehouse Modeling. *Journal of Convergence Information Technology* 6 (11): 48-53.

Chong, H.Y, & Zin, R.M (2012). Selection of dispute resolution methods: factor analysis approach. *Engineering Construction and Architectural Management* 19 (4): 428-443.

Chong, H.Y, & Phuah.Th. (2013). Incorporation of database approach into contractual issues: Methodology and practical guide for organizations. Automation in Construction 31: 149-157.

Chong, H.Y, Zin, R.M. &. Chong, S.C. (2013). Employing Data Warehousing for Contract Administration: e-Dispute Resolution Prototype. Journal of Construction Engineering and Management 139: 611-619.

CIB-TG57. Industrialisation in Construction. 18th CIB World Building Congress, (2010) Salford, United Kingdom.

CIB-TG62. Built Environment Complexity. 18th CIB World Building Congress, (2010) Salford, United Kingdom.

CIB-TG74. New Production and Business Models in Construction. (2010) Salford, United Kingdom.

CIB-W055. Building Economics. *In:* Barret, P. E. A., ed. 18th CIB World Building Congress, (2010) Salford, United Kingdom.

CIB-W078. Information Technology for Construction. *In:* Barret, P. E. A., ed. 18th CIB World Building Congress (2010) Salford, United Kingdom.

CIB-W092. Procurement Systems. *In:* Barret, P. E. A., ed. 18th CIB World Building Congress, (2010) Salford, United Kingdom.

CIB-W096. Architectural Management and Open Building Implementation. *In:* Barret, P. E. A., ed. 18th CIB World Building Congress, (2010) Salford, United Kingdom.

CIB-W102. Information and Knowledge Management in Building. *In:* Barret, P. E. A., ed. 18th CIB World Building Congress, (2010) Salford, United Kingdom.

CIB-W113. Law and Dispute Resolution. *In:* Barret, P. E. A., ed. 18th CIB World Building Congress, (2010) Salford, United Kingdom.

CIDB (2006). Construction Industry Master Plan 2006-2015. *Construction Industry Development Board (CIDB) Malaysia.*

CIDB (2008). Malaysia Construction Industry Annual Report. *In:* CIDB (ed.).

CIDB. (2011). *Building Materials Cost Index by Category of Building and Region* [Online]. Available: http://www.cidb.gov.my/v6/files/info/semenanjung2011_Jadual1.pdf.

CIDB. (2012a). *Number and Value of Projects Awarded By Category as of December (2011)* [Online]. Available: http://www.cidb.gov.my/v6/files/pub/BahagianKeduaQ42011.pdf [Accessed 21/01/2012 2012].

CIDB. (2012b). *Number and Value of Projects Awarded By Status of Contractors and Type of Contract as of December (2011)* [Online]. Available: http://www.cidb.gov.my/v6/files/pub/BahagianKeduaQ42011.pdf [Accessed 21/01/2012 2012].

CIDB. (2012c). *Number and Value of Projects Awarded By Type of Work and Type of Contract as of December (2011)* [Online]. Available: http://www.cidb.gov.my/v6/files/pub/BahagianKeduaQ42011.pdf [Accessed 21/01/2012 2012].

CIDB & CREAM (2011). A Case Study of BIM Implementation in Malaysia.

Clements, K. W., Lan, Y. & Seah, S. P. (2012). The Big Mac Index two decades on: an evaluation of burgernomics. *International Journal of Finance & Economics.*

Coates, P., Arayici, Y., Koskela, L., Kagioglou, M., Usher, C. & O'reilly, K. (2010). The key performance indicators of the BIM implementation process.

Computer-Integrated-Construction-Research-Program (2012). BIM Planning Guide for Facility Owners - Version 1.0, April. The Pennsylvania State University, University Park, PA, USA.

Constructech. (2012). *Big Data – How Does It Impact Construction Projects?* [Online]. Available: http://www.constructech.com/news/articles/article.aspx?article_id=9497 [Accessed October 2012.

Constructing-Excellence. (2004). *Supply Chain Management* [Online]. Available: http://www.constructingexcellence.org.uk/pdf/fact_sheet/supplychain.pdf 2012].

Constructing-Excellence. (2012). *What is Avanti?* [Online]. Available: http://www.constructingexcellence.org.uk/ceavanti/about.jsp.

Construction-Industry-Council (2013). Best Practice Guide For Professional Indemnity Insurance When Using Building Information Models. *1st Edition.* Construction Industry Council www.cic.org.uk.

Construction-Industry-Development-Board. (2011). *Regulations* [Online]. Available: http://www.cidb.gov.my/v6/?q=en/content/regulations.

Construction-Project-Information-Committee (2012). CPIC User Guides.

Construction-Research-Institute-Malaysia. (2011). *Industrialised Building System (Ibs): Towards Sustainable Implementation* [Online]. Available: http://ibsresearch.blogspot.com/2011_02_01_archive.html.

CREAM. (2012). *Integrated Design Delivery Solution (IDDS) / Building Information Modeling (BIM)* [Online]. Available: http://www.cream.com.my/index.php/r-a-d-and-consultancy/research-focus/integrated-design-delivery-solution.

Creswell, J. W. & Clark, V. L. P. (2007). *Designing and conducting mixed methods research*, Wiley Online Library.

Curado, C. (2006). The knowledge based-view of the firm: from theoretical origins to future implications. *Instituto Superior de Economia e Gestão–Departamento de Gestão Working paper series.*

Curado, C. & Bontis, N. (2006). The knowledge-based view of the firm and its theoretical precursor. *International Journal of Learning and Intellectual Capital,* Volume 3, 367-381.

Davis, F. D. (1985). *A technology acceptance model for empirically testing new end-user information systems: Theory and results.* Massachusetts Institute of Technology, Sloan School of Management.

Deraniyagala, S. U. (1996). Pre- and Protohistoric Settlement in Sri Lanka. In: A.B.A.C.O. S.R.L (ed.) International Union of Prehistoric and Protohistoric Sciences XIII Congress. Forli, Italy.

De Vaus, D. A. (1995). 'Administering questionnaires', Surveys in Social Research, 4th edn. *St Leonards: Allen & Unwin,* 106-125.

Deem, R. (1997). Focus groups. Lancaster University, Department of Educational Research,.

Denscombe, M. (2007). *The good research guide: for small-scale social research projects*, Open University Press.

De Silva, N. (2011). Promoting the facilities management profession in the project development phase of high-rise buildings in Sri Lanka. *Built-Environment Sri Lanka,* 9 -10, 37-44.

Dossick, C. S. & Neff, G. (2010). Organizational divisions in BIM-enabled commercial construction. *Journal of Construction Engineering and Management,* 136, 459.

Douma, S. W. & Schreuder, H. (2008). *Economic Approaches to Organizations*, Prentice Hall.

Dwyer, T. (2012). Further Dimensions. *CIBSE Journal,* January 2012.

Easterby, M., Thorpe, R. & Lowe, A. (1991). 'The ingredients of successful research'. *Chapter 2 of Management Research - An Introduction.* Sage, London.

Eastman, C., Fisher, D., Lafue, G., Lividini, J., Stoker, D. & Yessios, C. (1974). An Outline of the Building Description System. Pittsburgh: Institute of Physical Planning, Carnegie-Mellon University, Pittsburgh.

Eastman, C. M., Teicholz, P., Sacks, R. & Liston, K. (2011). *BIM handbook: a guide to building information modeling for owners, managers, designers, engineers, and contractors*, Wiley.

Eclipse-Research-Consultants (2005). Better designed buildings: improving the valuation of intangibles.

Egan, J. (1998). Rethinking Construction, Construction Task Force Report. *Department of the Environment, Transport and the Regions, London.*

Eisenhardt, K. M. & Santos, F. M. (2002). Knowledge-based view: A new theory of strategy. *Handbook of strategy and management,* 164.
Evans, H., Haryott, R., Haste, N., Jones, A. (1998). The Long Term Costs of Owning and Using Buildings, Royal Academy of Engineering, London.
Exactal. n.d. *Use CostX® to power your education* [Online]. Available: http://www.exactal.com/products/costx-educational [Accessed 20 Aug 2014].
Fairclough, S. J. 2002. Rethinking Construction Innovation and Research: A Review of Government R & D Policies and Practices, Department of Trade and Industry, London.
Faisol, N. (2010). *An investigation of relational contracting norms in construction projects in Malaysia.* PhD, Loughborough University.
Fathi, M. S., Abedi, M. & Mirasa, A. K. (2012a). Construction industry experience of industralised building system in Malaysia.
Fathi, M. S., Abedi, M. & Rawai, N. (2012b). The Potential of Cloud Computing Technology for Construction Collaboration. *Applied Mechanics and Materials,* 174, 1931-1934.
Fellows, R. (2011). Satisfying the Client? *In:* Preece, C. N., Fathi, S.M., Padfield, R., Papargyropoulou, E., Abdullah, M.M.B., Rogers, J. (ed.) *International Construction Business & Management Symposium.*
Fern, E. F. (2001). *Advanced focus group research*, Sage Publications, Inc.
Foss, N. J. (1996). Knowledge-based approaches to the theory of the firm: Some critical comments. *Organization Science*, 470-476.
Foss, N. J. (1997). *Resources, firms, and strategies: a reader in the resource-based perspective*, Oxford University Press, USA.
Foss, N. J. (2006). *Strategy, economic organization, and the knowledge economy: the coordination of firms and resources*, Oxford University Press, USA.
Fox, W. (2000). *Ethics and the built environment*, Routledge.
FreeCAD. (2012). Available: http://free-cad.sourceforge.net/.
Galabova, L. & Ahonen, G. (2011). Is intellectual capital-based strategy market-based or resource-based?: On sustainable strategy in a knowledge-based economy. *Journal of Human Resource Costing & Accounting,* 15, 313-327.
Gartner-Research. (2012). *Interpreting Technology Hype* [Online]. Available: http://www.gartner.com/technology/research/methodologies/hype-cycle.jsp.
GBCSL. (2014). Green Building Council of Sri Lanka [Online]. Available: http://srilankagbc.org/ [Accessed 8 Sep 2014].
George, H. (2012). *BIM collaboration and engineering – many heads are better than one* [Online]. Available: http://www.thenbs.com/topics/bim/articles/bimCollaborationAndEngineering.asp.
Golzarpoor, H. (2010). *Application of BIM in sustainability analysis.* Universiti Teknologi Malaysia.
GOM (2012). Federal Department of Town and Country Planning. *The Doctrine of Universal Planning and Development.*
Grant, R. M. (2002). The knowledge-based view of the firm. *The strategic management of intellectual capital and organizational knowledge*, 133-148.
Graphisoft. n.d.-a. *ArchiCAD Revolutionizes BIM Collaboration* [Online]. Graphisoft. Available: http://www.graphisoft.com/bim_server [Accessed 20 April 2014].
Graphisoft. n.d.-b. *MyArchiCAD* [Online]. Available: https://myarchicad.com [Accessed 20 Aug 2014].
Green-Building-Index. (2012). *Green Building Index* [Online]. Available: http://www.greenbuildingindex.org/.
Greene, J. C. & Caracelli, V. J. (1997). Advances in Mixed-Method Evaluation: The Challenges and Benefits of Integrating Diverse Paradigms: New Directions for Evaluation (Paperback)(Series: JB PE Single Issue (Program) Evaluation).
Grilo, A. & Jardim-Goncalves, R. (2010). Value proposition on interoperability of BIM and collaborative working environments. *Automation in Construction,* 19, 522-530.
Groves, R. M., Fowler, F. J., Couper, M. P., Lepkowski, J. M. & Singer, E. (2009). *Survey methodology*, John Wiley & Sons Inc.
Grundy, T. (2006). Rethinking and reinventing Michael Porter's five forces model. *Strategic Change,* 15, 213-229.
Gu, N. & London, K. (2010). Understanding and facilitating BIM adoption in the AEC industry. *Automation in Construction,* 19, 988-999.

Gu, N. & Singh, V. (2009). Building information modelling: implications for the AEC industry| NOVA. The University of Newcastle's Digital Repository.

Gu, N., Singh, V., Tsai, J., Taylor, C., London, K. & Brankovic, L. Industry perception of BIM adoption in design sector. (2008). International Islamic University Malaysia, Kulliyyah of Architecture and Environmental Design.

Guba, E. G. & Lincoln, Y. S. (1994). Competing paradigms in qualitative research. *Handbook of qualitative research,* 2, 163-194.

Gunasekara, K. & Jayasena, H. S. (2013). Identification of a Technological Framework for Implementing Building Information Modelling In Sri Lanka. *In:* Sandanayake, Y. G. & Fernando, N. G. (eds.) *The Second World Construction Symposium 2013: Socio-Economic Sustainability in Construction.* Colombo, Sri Lanka: Ceylon Institute of Builders.

Gunathilake, S. & Jayasena, H. S. Developing a relational approach to contracting: theSri Lankacontext. Proceedings of the CIB International Conference on Building Education and Research, (2008) Kandalama, Sri Lanka. CIB, 280-281.

Hamid, A., Hamidah, S., Takim, R. & Nawawi, A. H. An integrated value management (IVM) for construction projects in Malaysia. (2011). IEEE, 421-425.

Hardin, B. (2009). *BIM and construction management: proven tools, methods, and workflows*, Wiley Pub.

Harris, F., Mccaffer, R. & Edum-Fotwe, F. (2006). *Modern construction management*, Wiley-Blackwell.

Hofstede, G. H., Hofstede, G. J. & Minkov, M. (2010). *Cultures and organizations: software for the mind,* McGraw-Hill Professional.

Hok-Network; Macleamy, P. (2010). The Future of the Building Industry (5/5): BIM, BAM, BOOM!

Hughes, W., Ancell, D., Gruneberg, S. & Hirst, L. (2004). Exposing the myth of the 1: 5: 200 ratio relating initial cost, maintenance and staffing costs of office buildings. 2004. 1-3.

Huovinen, P. (2010a). Enhancement of Business Management for a Better Construction World: A Review of a 62-Concept Population Published Between 1990-2009. 9.

Huovinen, P. (2010b). Viable Concepts for Managing Global Businesses in a Better Construction World. 21.

ICTAD. (2014). Welcome to Institute for Construction Training and Development [Online]. Available: http://www.ictad.lk/ [Accessed 7 Sep 2014].

Ibrahim, A. R. B., Roy, M. H., Ahmed, Z. & Imtiaz, G. (2010a). An investigation of the status of the Malaysian construction industry. *Benchmarking: An International Journal,* 17, 294-308.

Ibrahim, A. R. B., Roy, M. H., Ahmed, Z. U. & Imtiaz, G. (2010b). Analyzing the dynamics of the global construction industry: past, present and future. *Benchmarking: An International Journal,* 17, 232-252.

Ibrahim, R. (2011). Demystifying the Arduous Doctoral Journey: The Eagle Vision of a Research Proposal. *Electronic Journal of Business Research Methods,* 9, 130-140.

IEM (2011). The Institution of Engineers, Malaysia - Constitution and By-Laws.

IEM (2012). Annual Report. Kuala Lumpur: The Institution of Engineers, Malaysia.

IMF (2012). World Economic outlook - April 2012. Washington, DC: International Monetary Fund.

Indiana-University (2012). Building Information Modeling (BIM) Guidelines and Standards for Architects, Engineers, and Contractors.

Information And Communication Technology Agency. (2014). *Sri Lanka soars in eGovernment* [Online]. Information and Communication Technology Agency,. Available: http://www.icta.lk/en/icta/90-general/1436-sri-lanka-soars-in-egovernment.html [Accessed 20 Aug 2014].

Isikdag, U., Underwood, J. & Aouad, G. (2008). An Investigation into the Applicability of Building Information Models in Geospatial Environment in support of Site Selection and Fire Response Management Processes. *Advanced Engineering Informatics,* 22(4), pp.504-519.

Ismail, S., Yusof, A. M. & Han, W. S. (2012). Elements of Relational Contract in the Delivery of Public Infrastructure in Malaysia. *IBIMA Business Review,* 2012.

Jaafar, M., Abdul Aziz, A. R., Ramayah, T. & Saad, B. (2007a). Integrating information technology in the construction industry: Technology readiness assessment of Malaysian contractors. *International Journal of Project Management,* 25, 115-120.

Jaafar, M., Ramayah, T., Abdul-Aziz, A. R. & Saad, B. (2007b). Technology readiness among managers of Malaysian construction firms. *Engineering, Construction and Architectural Management,* 14, 180-191.

Jaapar, A., Endut, I. R., Ahmad Bari, N. A. & Takim, R. (2009). The Impact of Value Management Implementation in Malaysia. *Journal of Sustainable Development,* 2, P210.

Jaapar, A., Takim, R., Ahmad Bari, N. A. & Endut, I. R. (2008). Value management guidelines for the Malaysian construction industry. *ALAM CIPTA, International Journal on Sustainable Tropical Design Research & Practice,* 3, 91-98.

Jayasena, H. S. & De Silva, R. (2013). Building cost information modelling framework. International Quantity Surveying BIM Conference. Hong Kong Polytechnic University, Hong Kong: The Hong Kong Institute of Surveyors.

Jayasena, H. S. & Uhanowitage, R. (2008). The Effect of Winner's Curse on Post-Contract Management. International Conference on Building Education and Research, Building Resilience, 2008 Kandalama, Sri Lanka. CIB, 284-285.

Jayasena, H. S. & Weddikkara, C. (2013). Assessing the BIM Maturity in a BIM Infant Industry. *In:* Sandanayake, Y. G. & Fernando, N. G. (eds.) (2013). *The Second World Construction Symposium 2013: Socio-Economic Sustainability in Construction.* Colombo, Sri Lanka: Ceylon Institute of Builders.

Jayasena, H. S. & Weddikkara, C. (2014). BIM Software Environment for Projects in Sri Lanka. *In:* Sandanayake, Y. G., Fernando, N. G. & Karunasena, G. I. (eds.) *The 3rd World Construction Symposium.* Colombo: Ceylon Institute of Builders - Sri Lanka.

Jayasena, S. (2014). Expanding It Related Career Opportunities With Advent Of Bim. *Edex Magazine.*

Jernigan, F. E. (2008). *Big Bim Little Bim - The Practical Approach To Building Information Modeling : integrated practice done the right way,* Salisbury, USA, 4Site Press.

Johnson, R. B. & Onwuegbuzie, A. J. (2004). Mixed methods research: A research paradigm whose time has come. *Educational researcher,* 33, 14-26.

Joseph, A. L. & Jayasena, H. S. Impediments to the Development of Design and Build Procurement System in Sri Lanka. International Conference on Building Education and Research, Building Resilience, (2008) Kandalama, Sri Lanka. CIB, 286-287.

Kala, T. E. A. (2010). Case Study using an integrated 5D & location based planning system in a large hospital construction project. *Lean Construction,* 2010, 102-112.

Kamar, K. A. M. (2011a). *Industrialised Building System Research and Information* [Online]. Available: http://ibsresearch.blogspot.com/2012/01/8th-ibs-rountable-mechanisation-through.html [Accessed 21/01/2012 2012].

Kamar, K. A. M., And Hamid, Z.A. (CIDB-CREAM) (2011b). Supply Chain Strategy for Contractor in Adopting Industrialised Building System (IBS). *Australian Journal of Basic and Applied Sciences,* 5, 2552-2557.

Kaplan, S., Schenkel, A., Von Krogh, G. & Weber, C. (2001). Knowledge-based theories of the firm in strategic management: a review and extension. *Academy of Management Review.*

Kareem, H. I. A. & Bakar, A. H. A. (2011). Identifying IT benefits for Malaysian construction companies. ITcon.

Kelly, J. & Male, S. (1993). *Value Management in Design and Construction: the economic management of projects*, Taylor & Francis.

Kelly, J., Male, S. & Graham, D. (2004). *Value management of construction projects*, Wiley Online Library.

Kent, D. C. & Becerik-Gerber, B. (2010). Understanding construction industry experience and attitudes toward integrated project delivery. *Journal of Construction Engineering and Management,* 136, 815.

Khanzode, A. (2010). An integrated, virtual design and construction and lean (IVL) method for coordination of MEP. *Unpublished Technical Report,* 187.

Khemlani, L. (2012). *Around the World with BIM* [Online]. AECbytes. Available: http://www.aecbytes.com/feature/2012/Global-BIM.html [Accessed 19 May 2013].

Kim, W. C. & Mauborgne, R. (2005). *Blue ocean strategy: How to create uncontested market space and make competition irrelevant*, Harvard Business Press.

Kitzinger, J. (1994). The methodology of focus groups: the importance of interaction between research participants. *Sociology of health & illness,* 16, 103-121.

Kitzinger, J. (1995). Qualitative research: introducing focus groups. *Bmj,* 311, 299-302.

Kitzinger, J. & Barbour, R. S. (1999). *Developing focus group research: politics, theory and practice*, Sage Publications Ltd.

Kogut, B. & Zander, U. (1996). What firms do? Coordination, identity, and learning. *Organization Science*, 502-518.

Koskela, L. (1992). Application of the new production philosophy to construction. Center for Integrated Facility Engineering - Stanford University.

Koskela, L. & Huovila, P. (1997). On Foundations of Concurrent Engineering. *Proc. 1st International Conference on Concurrent Engineering in Construction,* The Institution of Structural Engineers, London, 22-32.

Koskela, L. J., Howell, G. (2002). The underlying theory of project management is obsolete. PMI.

Krueger, R. A. & Casey, M. A. (2009). *Focus groups: A practical guide for applied research*, Sage.

Kuhn, T. S. (1996). *The structure of scientific revolutions*, University of Chicago press.

Laiserin, J. (2003). Graphisoft on BIM. *Laiserin Letter* [Online]. Available: http://www.laiserin.com/features/issue19/feature01.php [Accessed January 20].

Langston, C. & Best, R. (2013). Using the Big Mac index for comparing construction costs internationally. Conference Proceedings: The Queensland University of Technology Research Week International Conference, 2012. Queensland University of Technology, 1-11.

Latham, M. (1994). Constructing The Team: Joint Review of Procurement and Contractual Arrangements in the United Kingdom Construction Industry; final report.

Leahey, E. (2007). Convergence and confidentiality? Limits to the implementation of mixed methodology. *Social Science Research,* 36, 149-158.

Leedy, P. D. & Ormrod, J. E. (2005). *Practical Research: Planning and Design, 8th Edition*, Pearson Merrill Prentice Hall.

Liebich, T., Adachi, Y., Forester, J., Hyvarinen, J., Richter, S., Chipman, T., Weise, M. & Wix, J. (2013). *IFC4 Official Release* [Online]. buildingSMART International. Available: http://www.buildingsmart-tech.org/ifc/IFC4/final/html/index.htm [Accessed 01 Dec 2013].

Li, H., Huang, T., Kong, C., Guo, H., Baldwin, A., Chan, N. & Wong, J. (2008). Integrating design and construction through virtual prototyping. *Automation in Construction,* 17, 915-922.

Liao, C., Palvia, P. & Chen, J.-L. (2009). Information technology adoption behavior life cycle: Toward a Technology Continuance Theory (TCT). *International Journal of Information Management,* 29, 309-320.

Lim, V. A. J. (2008). *Lean construction: knowledge and barriers in implementing into Malaysia construction industry.* Universiti Teknologi Malaysia.

Lincoln, Y. S. & Denzin, N. K. (2003). *Turning points in qualitative research: Tying knots in a handkerchief,* Altamira Pr.

Lu, W. & Li, H. (2011). Building information modeling and changing construction practices. *Automation in Construction,* 20, 99-100.

Luthra, A. (2010). *Implementation of Building Information Modeling in Architectural Firms in India.* Master of Science, Purdue University.

Macleamy, P. Keynote. (2007) buildingSMART alliance™ National Conference, 2007.

Malaysia-Gov (2000). The Third Outline Perspective Plan. Putrajaya.

Malaysia-Gov (2007). Strategic Recommendation For Improving Environmental Practices In Construction Industry. *In:* (Cidb), C. I. D. B. (Ed.).

Malaysia-Gov (2009). Integration Of The Construction Industry Through Partnering – The Malaysian Initiative. *Integration Of The Value Chain In The Construction Industry,* 2012.

Malaysia-Gov (2013). Malaysia Population Clock. *In:* Department Of Statistics, M. (Ed.). Putrajaya.

Malaysia-Government. (2011). *Pemandu* [Online]. Available: http://etp.pemandu.gov.my/default.aspx.

Malaysian-Bar-Council. (2013). *Construction court on the cards* [Online]. Available: http://www.malaysianbar.org.my/legal/general_news/construction_court_on_the_cards.html [Accessed 16 February 2013].

Male, S. (2010). Improving Scotland`s Public Sector Estate. *Asset 2010.* Hilton Edinburgh Grosvenor Hotel: Institute for Resilient Infrastructure, School of Civil Engineering, University of Leeds.

Male, S., Kelly, J., Gronqvist, M. & Graham, D. (2005). Reappraising value methodologies in construction for achieving best value. 4-8.

Malhotra, N. (2010). *Marketing Research : An Applied Orientation, 6th edn,* Upper Saddle River, New Jersey, Pearson Education.

Managan, K., Araya, M., Layke, J. (2011). Driving Transformation To Energy Efficient Buildings: Policies and Actions. Institute for Building Efficiency, Business Council for Sustainable Energy, Center for Clean Air Policy, U.S. Green Building Council, and World Green Building Council.

Manning, M. & Mcmurray, D. (2009). Quantitative Research Methods. *In:* Mcmurray, D. (Ed.) *Study Guide Edu03263*. Lisamore, Australia: Southern Cross University.
Manning, M. & Munro, D. (2007). *The Survey Researcher's Spss Cookbook*, Pearson Education Australia.
MASTERSPEC (2012). New Zealand National BIM Survey 2012. *In:* HUBER, R. (ed.). Construction Information Ltd, PO Box 108 214, Auckland 1150, New Zealand
www.masterspec.co.nz
Matthews, O. & Howell, G. A. (2005). Integrated project delivery: an example of relational contracting. *Lean Construction Journal,* 2, 46-61.
Mcdonough-School-Of-Business. (2012). Five Forces - Critique. Georgetown-University.
Mcgraw-Hill-Construction (2009). Smartmarket Report: The Business Value of BIM.
Mcgraw-Hill-Construction (2010). Smart Market Report - The Business Value of BIM in Europe.
Mcintyre, L. (1999). The Practical Skeptic: Core Concepts in Sociology. *In:* Tichenor, V. (ed.) *Teaching Sociology.*
Mcmurray, D. (2009). Qualitative Research Methods. *Study Guide EDU03262.* Lismore, Australia: Southern Cross University.
Meikle, J. (2011). Modern Construction Economics: Theory and Application. *Construction Management and Economics,* 29, 765-766.
Meng, J., Elliott, K. M. & Hall, M. C. (2010). Technology Readiness Index (TRI): Assessing Cross-Cultural Validity. *Journal of International Consumer Marketing,* 22, 19-31.
Michigan-State-University (2008). Lean Construction – A Promising Future for MSU
Miller, D. C. & Salkind, N. J. (2002). *Handbook of research design and social measurement*, Sage Publications, Incorporated.
Mills, A. (2001). A systematic approach to risk management for construction. *Structural Survey,* 19, 245-252.
Ministry-of-Finance-Malaysia. (2011). *Quarterly Update on the Malaysian Economy – 3rd Quarter 2011* [Online]. Available: http://www.treasury.gov.my/pdf/ekonomi/sukutahun3_2011.pdf.
Ministry-Of-Finance-Malaysia (2012). Quarterly Update on the Malaysian Economy – 3rd Quarter 2012.
Mintzberg, H. (1994). *The rise and fall of strategic planning: Reconceiving roles for planning, plans, planners*, Free Pr.
Mitchell, M. & Jolley, J. (1992). Generating the research hypothesis: Tapping intuition, theory and existing research'. *Research Design Explained, 2nd edn.*, 37-90.
Morgan, D. L. (1997). *Focus groups as qualitative research*, Sage Publications, Inc.
Morgan, D. L. (1998). Practical strategies for combining qualitative and quantitative methods: Applications to health research. *Qualitative health research,* 8, 362-376.
Mossman, A. (2009). Why isn't the UK construction industry going lean with gusto? *Lean Construction Journal,* 5, 24-36.
Mowery, D. C., Oxley, J. E. & Silverman, B. S. (1998). Technological overlap and interfirm cooperation: implications for the resource-based view of the firm. *Research Policy,* 27, 507-523.
Muniandy, J. K. & Barnes, L. (2010). The link between Corporate Social Performance and Institutional Investors' Shareholdings in Malaysian Public Listed Companies. *International Review of Business Research Papers,* 6, 246-261.
Nachmias, D. & Nachmias, C. F. (2008). *Research methods in the social sciences: Seventh Edition*, Worth Publishers New York.
Narayanan, V. & Fahey, L. (2005). The relevance of the institutional underpinnings of Porter's five forces framework to emerging economies: An epistemological analysis. *Journal of Management Studies,* 42, 207-223.
National-Institute-Building-Sciences-And-Buildingsmart-Alliance (2012). The Interactive BIM Capability Maturity Model. *BIM_CMM_v1.9 (3).xls.* USA.
National-Institute-Building-Sciences (2007). National Building Information Modeling Standard (NBIMS) *3 Version 1.0 – Part 1: Overview, Principles, and Methodology.*
National-Institute-Building-Sciences (2012a). National Building Information Model Standard.
National-Institute-Building-Sciences (2012b). National Building Information Model Standard - Minimum BIM.
National-Institute-Of-Building-Sciences. (2012). *buildingSMART alliance to Collaborate with UK, Ireland on BIM Standards* [Online]. Available: http://www.nibs.org/news/100242/ [Accessed 04/07/2012.
Nelson, R. R. & Winter, S. G. (1982). *An evolutionary theory of economic change*, Belknap press.

NEUMAN, W. L. (2006). *Social Research Methods - Qualitative and Quantitative Approaches,* Boston, Pearson Education Inc.
New-York-City-Department-Of-Design-&-Construction (2012). BIM Guidelines. New York, USA.
NHMRC, A. C. C. (2007). National Statement on Ethical Conduct in Human Research. *NHMRC, Canberra.*
Nielsen. (2011). *The Nielsen Global Online Environmental Survey - Sustainable Efforts & Environmental Concerns* [Online]. Available: http://www.nielsen.com/us/en/insights/reports-downloads/2011/sustainable-efforts-environmental-concerns.html.
Nizarudin, N. D., Hussain, M., Ramzi, M. & Tukiman, I. (2010). The application of the Green Building Index (GBI) on sustainable site planning and management for residential new construction: prospects and future benefits.
Nonaka, I., Toyama, R. & Nagata, A. (2000). A firm as a knowledge-creating entity: a new perspective on the theory of the firm. *Industrial and corporate change,* 9, 1-20.
Nonaka, I., Von Krogh, G. & Voelpel, S. (2006). Organizational knowledge creation theory: Evolutionary paths and future advances. *Organization studies,* 27, 1179-1208.
Ofori, G. Challenges of construction industries in developing countries: lessons from various countries. (2000). Citeseer, 15-17.
Ofori, G. (2007). Construction in developing countries. *Construction Management and Economics,* 25, 1-6.
Ofori, G. (2011). *RE: email to author.*
Ofori, G. & Lewis, T. Construction and the environment in developing countries. (1993). Port of Spain, Trinidad, 15-22.
Olantunjii, O. A., (2011), 'A preliminary review on the legal implications of BIM and model ownership', *ITcon* 16, 687-696.
PAM. (2012). *Calendar, Events and Seminars* [Online]. Available: http://www.pam.org.my/images/events/2012/PAM_Events_Calendar_Jan_-_Dec.pdf.
Parasuraman, A. (2000). Technology readiness index (TRI). *Journal of Service Research,* 2, 307.
Perry, C. (1998). A Structured Approach to Presenting Theses. *Notes*, 1-57.
Perry, C. (2002). A Structured Approach to Presenting Theses: Notes for Students and their Supervisors.
Pheng, L. S. & Shang, G. (2011). The Application of the Just-in-Time Philosophy in the Chinese Construction Industry. *Journal of Construction in Developing Countries,* 16, 91-111.
Pickavance, K. (2013). *The CIOB Complex Projects Contract 2013* [Online]. NBS. Available: http://www.thenbs.com/topics/contractslaw/articles/theCIOBComplexProjectsContract2013.asp [Accessed 21 Aug 2014].
Polanyi, M. & Sen, A. (1983). *The tacit dimension*, Peter Smith Gloucester, MA.
Porter, M. E. (1998a). *Competitive advantage: creating and sustaining superior performance: with a new introduction*, Free Pr.
Porter, M. E. (1998b). *Competitive strategy: techniques for analyzing industries and competitors: with a new introduction*, Free Pr.
Porter, M. E. (2008). *On Competition*, Harvard Business Press.
Porter, M. E. & Kramer, M. R. (2006). Strategy and Society: The Link Between Competitive Advantage and Corporate Social Responsibility. *Harvard Business Review*, 78-92.
Porter, M. E. & Millar, V. E. (1985). How information gives you competitive advantage. *Harvard Business Review,* 63, 149-160.
Post, N. M. (2011). An Unprecedented 11 Partners Propel Integrated Project Delivery at Sutter's New California Hospital. *Engineering News Record.* ENR September 19, 2011 ed. New York: McGraw-Hill Construction.
Ramasamy, B., Ting, H. W. & Yeung, M. C. Y. (2007). Does It Pay To Be Good In Developing Countries? The Relationship Between Csr And Financial Performance In Malaysia. *Asian Academy Of Management Journal Of Accounting And Finance,* Vol. 3, 21–36.
Razak, A., Roy, M.H., Ahmed, Z., Imtiaz, G. (2010). An investigation of the status of the Malaysian construction industry. *Benchmarking: An International Journal of Asian Economics,* Vol. 17 294-308.
Redmond, A.; Hore, A.; Alshawi, M.; West, R., (2012), 'Exploring how information exchanges can be enhanced through Cloud BIM', *Automation in Construction 24, 175-183.*
REHDA. (2011). *Real Estate And Housing Developers' Association Malaysia-Towards Sustainable Development.* [Online]. Available: http://www.rehda.com/bulletin/10/index.html [Accessed 26/07/2011 2011].

Riazi, S. R. M., Skitmore, M., Cheung, Y.K.F. (2011). The use of supply chain management to reduce delays: Malaysian public secto construction projects.

Robinson, N. M., Lavers, A. P., Tan, G. & Chan, R. (1988). *Construction Law in Singapore and Malaysia*, Butterworths.

Robson, C. (2002). *Real world research: A resource for social scientists and practitioner-researchers*, Wiley-Blackwell.

Rogers, E. M. (1995). *Diffusion of innovations*, Free Pr.

Rollinson, D. (2008). *Organisational Behaviour and Analysis*, Prentice Hall.

Roper, K. BIM Implementation: PESTEL Drivers & Barriers (Cross-national Analysis). (2012). Ifma.

Royal-Institute-Of-British-Architects. (2007). *Outline Plan of Work 2007* [Online]. Available: http://www.architecture.com/Files/RIBAProfessionalServices/Practice/OutlinePlanofWork(revised).pdf.

Royal-Institute-Of-British-Architects. (2013). *Outline Plan of Work 2013* [Online]. Available: http://www.architecture.com/TheRIBA/AboutUs/Professionalsupport/RIBAOutlinePlanofWork2013.aspx#.UkO9QIbQkgM.

Royal-Institute-Of-Chartered-Surveyors (2011). The construction supply chain.

Ruddock, L. (2007). The Economic Value of Construction - Achieving a Better Understanding *The construction and building research conference of the Royal Institution of Chartered Surveyors.* Georgia Tech, Atlanta USA.

Sacks, R. & Barak, R. (2008). Impact of three-dimensional parametric modeling of buildings on productivity in structural engineering practice. *Automation in Construction,* 17, 439-449.

Sacks, R., Kaner, I., Eastman, C. M. & Jeong, Y.-S. (2010a). The Rosewood experiment—Building information modeling and interoperability for architectural precast facades. *Automation in Construction,* 19, 419-432.

Sacks, R., Koskela, L., Dave, B. A. & Owen, R. (2010b). Interaction of Lean and Building Information Modeling in Construction. *Journal of Construction Engineering and Management,* 136, 968.

Saifulnizam, M., Coffey, V., Preece, C. (2010). Marketing VM services to achieve competitive advantage in the Malaysian Construction Sector. *Management in Construction Researchers Association (MiCRA) 2010.* Universiti Teknologi MARA, Shah Alam, Malaysia.

Samuelson, O. & Björk, B. C. (2011). Adoption processes for EDM, EDI and BIM technologies in the construction industry.

Saxon, R. (2005). Be valuable: a guide to creating value in the built environment. *Constructing Excellence, London.*

Schumpeter, J. A. (1994) [1942]. Capitalism, Socialism and Democracy. London: Routledge.

Sebastian, R., *et al.* (INPRO) (2011). Strategy For Implementation - lifecycle and BIM system.

Sekaran, U. (2000). 'Scientific investigation', Research Methods for Business: A Skill Building Approach. 3rd Edn. New York: John Wiley & Sons, 19–35.

Senescu, R. & Haymaker, J. (2009). Improving Design Processes through Collaborating, Sharing, and Understanding. CIFE Working Paper 2009-2010 Projects, Stanford University.

Shahrin, F., Johansen, E., Lockley, S. & Udeaja, C. (2010). Effective Capture, Translating And Delivering Client Requirements Using Building Information Modelling (BIM) Technology. ARCOM Research Workshop On Decision-Making Across Levels, Time And Space: Exploring Theories, Methods, (2010). 38.

Shou, W., Wang, J., Wang, X., & Chong, HY. (2014). A Comparative Review of Building Information Modelling Implementation in Building and Infrastructure Industries. *Archives of Computational Methods in Engineering*, 1-18. DOI 10.1007/s11831-014-9125-9

Singapore-Gov (2010). Building and Construction Authority - CORENET BIM Team - BIM Submission Guideline for Architectural Discipline.

Singapore-Gov. (2011). *Build Smart* [Online]. Available: http://www.bca.gov.sg/Publications/BuildSmart/others/buildsmart_11issue9.pdf.

Singapore-Gov (2012a). BIM Deployment Plan. *In:* BUILDING-AND-CONSTRUCTION-AUTHORITY (Ed.). Singapore.

Singapore-Gov (2012b). Singapore BIM Guide. *In:* Authority, B. C. (ed.). Singapore.

Singapore-Gov. (2013). *Technology Adoption: Building Information Model (BIM) Fund (Enhanced)* [Online]. Available: http://www.bca.gov.sg/BIM/bimfund.html.

Singh, V., Gu, N. & Wang, X. (2011). A theoretical framework of a BIM-based multi-disciplinary collaboration platform. *Automation in Construction,* 20, 134-144.

Slack, N., Chambers, S. & Johnston, R. (2010). *Operations Management*, Financial Times/Prentice Hall.

Smith, P. (2014). BIM & the 5D Project Cost Manager. *Procedia - Social and Behavioral Sciences,* 119, 475–484.

Snook, K. (2011). *Building Information Modelling* [Online]. Available: http://www.cpic.org.uk/en/bim/building-information-modelling.cfm [Accessed 08/04/(2012).

Spender, J. C. (1996). Making knowledge the basis of a dynamic theory of the firm. *Strategic Management Journal,* 17, 45-62.

Stake, R. E. (1995). *The art of case study research*, Sage Publications, Inc.

Stewart, D. W., Shamdasani, P. N. & Rook, D. W. (2007). *Focus groups: Theory and practice*, Sage Publications, Inc.

Succar, B. (2009). Building information modelling framework: A research and delivery foundation for industry stakeholders. *Automation in Construction,* 18, 357-375.

Succar, B. The five components of BIM performance measurement. (2010).

Sawhney, A. & Maheswari, J. U., (2013), 'Design Coordination Using Cloud-based Smart Building Element Models', *International Journal of Computer Information Systems and Industrial Management Applications* 5, 445- 453

Taib, M. S. & Ho, C. S. (2008). Planning system in Malaysia.

Taib, T. (2010). *Efficiency in construction process.* Masters, Universiti Teknologi Malaysia

Tan, C. B. (2011). *Level Of Awareness Towards Building Information Modelling (Bim) Among Quantity Surveyors In Malaysia.* Bachelors, Universiti Teknologi Malaysia.

Tan, W. & Suranga, H. (2008). Winner's Curse in the Sri Lankan Construction Industry. The International Journal of Construction Management, 8, 29-35.

Tashakkori, A. & Teddlie, C. (1998). *Mixed methodology: Combining qualitative and quantitative approaches*, Sage Publications, Inc.

Tashakkori, A. & Teddlie, C. (2003). *Handbook of mixed methods in social & behavioral research*, Sage Publications, Inc.

Tek, O. H. Quality and value management in construction - Malaysian Experience. (2004).

Tekla. n.d. *Learn Building Information Modeling for Free* [Online]. Available: http://www.tekla.com/solutions/students.

Tekla. (2012). *Home* [Online]. Available: www.tekla.com.my [Accessed 09.07.2012.

Teo, X. Q. (2012). *A study of building information modeling (BIM) in Malaysia construction industry.* UTAR.

The-Economic-Planning-Unit (2010). Tenth Malaysia Plan 2011-2015. *In:* Department, P. M. S. (ed.). Putrajaya.

The-Economist (2013). Big Mac Index.

The-NBS (2012a). National BIM Report 2012.

The-NBS (2012b). NBS BIM Research Report *BIM Roundtable Discussion.*

The-NBS. (2013a). *Building Information Modelling (BIM)* [Online]. Available: http://www.thenbs.com/topics/bim/articles/bimInConstruction.asp.

The-NBS. (2013b). *Create - Specification Tool* [Online]. Available: http://www.thenbs.com/products/nbsCreate/index.asp.

Thomsen, C. & Faia, F. (2010). Managing integrated project delivery. *CMAA. http:/Icmaanet. org/files/shared/ng_Integrated_Project_Delivery_1*, 1-19.

Ticehurst, G. W. & Veal, A. J. (2000). 'Questionnaire surveys'. *Business Research Methods: A Managerial Approach*, 135-158.

Treasury, H. M. (2008). Infrastructure procurement: delivering long-term value. *UK Government.*

Treasury, H. M. (2011). Appraisal and evaluation in central government (The Green Book). *London: The Stationery Office.*

U.S.A-Gov, C. I. A.-. (2011). *The World Factbook* [Online]. Available: https://www.cia.gov/library/publications/the-world-factbook/geos/my.html.

UK-Gov (2011a). Building Information Modelling (BIM) Working Party Strategy Paper. Department of Business, Innovation and Skills

Constructing Excellence.

UK-Gov (2011b). Government Construction Strategy. *In:* OFFICE, C. (ed.). London.
Underwood, J. & Isikdag, U. (2009). *Handbook of research on building information modeling and construction informatics: concepts and technologies*, IGI-Global.
Underwood, J., Isikdag, U., Kuruoglu, M. & Irtem, E. (2011). Enabling The Strategic Role of ICT in the Construction Industry: A Comparative Analysis of the UK and Turkish Sectors. *in: Advances in Business and Management,* Nova Science Publishers Inc., New York, USA.
United-Nations (1987). Our Common Future (The Brundtland Report).
United-Nations (2001). International Recommendations for Construction Statistics. New York.
Van Berlo, L. A. H. M., Beetz, J., Bos, P., Hendriks, H. & Van Tongeren, R. C. J. (2012). Collaborative engineering with IFC; new insights and technology. *9th European Conference on Product and Process Modelling.* Iceland: Innovation Center Iceland.
Van Horen, B. & Pinnawala, S. (2006). Sri Lanka. In: ROBERTS, B. & KANALEY, T. (eds.) Urbanization and Sustainability in Asia: Case Studies of Good Practice. Philippines: Asian Development Bank.
Venkatesh, V., Morris, M. G., Davis, G. B. & Davis, F. D. (2003). User acceptance of information technology: Toward a unified view. *Mis Quarterly***,** 425-478.
Vico Software. n.d. *Student Care Package* [Online]. Available: http://www.vicosoftware.com/student-care-package/tabid/298610/ [Accessed 20 Aug 2014].
Von Krogh, G., Nonaka, I. & Aben, M. (2001). Making the most of your company's knowledge: a strategic framework. *Long range planning,* 34, 421-439.
Vrijhoef, R. & Koskela, L. (2000). The four roles of supply chain management in construction. *European Journal of Purchasing & Supply Management,* 6, 169-178.
Wang, J. (2008). *Integrated Project Delivery.* Masters, University of Cincinnati.
Wernerfelt, B. (1984). A resource based view of the firm. *Strategic Management Journal,* 5, 171-180.
Whelton, M. & Ballard, G. Wicked problems in project definition. (2002).
Wijayakumar, M. & Jayasena, H. S. (2013). Automation of BIM Quantity Take-Off to Suit QS's Requirements. *In:* Sandanayake, Y. G. & Fernando, N. G. (eds.) (2013). *The Second World Construction Symposium 2013: Socio-Economic Sustainability in Construction.* Colombo, Sri Lanka: Ceylon Institute of Builders.
Wild, A. (2001). Construction projects as teams or situations: Criticising the techniques of communication approach. *In:* Akintoye, A., ed. Proceedings of the Seventeenth Annual Conference of the Association of Researchers in Construction Management, 5-7 Sep 2001 2001 University of Salford, Manchester, UK. 495-504.
Winch, G. (2003). Models of manufacturing and the construction process: the genesis of re-engineering construction. *Building Research & Information,* 31, 107-118.
Won, J. & Lee, G. (2010). Identifying the Consideration Factors for Successful BIM Projects. *In:* Tizani, W. (ed.) *the International Conference on Computing in Civil and Building Engineering.* Nottingham: Nottingham University Press.
Wong, W. W. (2012). *The internationalization of Malaysian engineering consulting services firms.* Doctoral, Southern Cross University.
World Intellectual Property Organization. n.d. *What is Intellectual Property?* [Online]. Available: http://www.wipo.int/about-ip/en/ [Accessed 21 Aug 2014].
Yam, S., Ismail, M. & Yin, T. S. (2008). Corporate social responsibility in Malaysia housing development-the developer's perspective.
Yin, R. K. (2009). *Case study research: Design and methods*, Sage publications, INC.
Yin, X., Wonka, P. & Razdan, A. (2009). Generating 3d building models from architectural drawings: A survey. *Computer Graphics and Applications, IEEE,* 29, 20-30.
Young, N. E. A. (2009). The Business Value of BIM: Getting Building Information Modeling to the Bottom Line. *Smart Market Report.* New York, NY: McGraw Hill Construction.
Zikmund, W. G. (2003a). *Business Research Methods,* Ohio, Cengage Learning.
Zikmund, W. G. (2003b). 'Sample designs and sampling procedures', Business Research Methods, 7th edn. *Ohio: South-Western Thomson,* 368-400.
Zikmund, W. G., Babin, B. J., Carr, J. C. & Griffin, M. (2010). *Business Research Methods*, South-Western Pub.

Subject Index

A

B

John Rogers, Heap-Yih Chong, Christopher Preece, Chai Chai Lim and Himal Suranga Jayasena

Q

R

S

T

V